Lecture Notes in Geoinformation and Cartography

For further volumes:
http://www.springer.com/series/7418

Susanne Rau · Ekkehard Schönherr
Editors

Mapping Spatial Relations, Their Perceptions and Dynamics

The City Today and in the Past

Springer

Editors
Susanne Rau
Ekkehard Schönherr
Philosophische Fakultät
Universität Erfurt
Erfurt
Germany

Additional material to this book can be downloaded from http://extras.springer.com.

ISBN 978-3-319-37608-0 ISBN 978-3-319-00993-3 (eBook)
DOI 10.1007/978-3-319-00993-3
Springer Cham Heidelberg New York Dordrecht London

Printed on acid-free paper

Springer is part of Springer Science+Business Media (www.springer.com)

Preface

Cities have been producing their own maps for purposes such as representation, control or orientation, right from the beginning of the fifteenth century, if not before. The introduction of geo-information systems and Google-maps has given rise to the firm belief that it is possible to represent, in an objective way, our spatial environment, especially with regard to the cities we live in. While mapping and representation technologies have been constantly changing, the results have mostly remained the same, i.e., static and objective (or misleadingly objective) images of a city. These kinds of maps would seem to fail to take two major aspects into consideration: first, the evolution of urban form for which a dynamic map, a film, or a dynamic 3D-model might provide more adequate means of representation and second, socio-spatial relations, i.e., the significance of forms and places, including perceptions of cities (as well as their perceived changes) which will always differ according to social, gender-based, internal/external factors, or other points of view.

Current research into historical and cultural studies on people's perceptions of sites and spatial arrangements, together with their discourses and interpretations, has proved to be essential for a deeper understanding of cities. If we take these findings into consideration, i.e., concerning the diversity and evolution not only of urban morphology itself, but also of spatial relations and people's perceptions of these relations, the following question arises: how should we design maps? Are geo-information systems the road ahead or are they a cul-de-sac?

The volume's focus is, therefore, not on the mapping of (static) urban space (past and present) but on the following topics: the potential mapping of perceived urban space, spatial hierarchies as a consequence of social usages (by a variety of active participants) and spatio-temporal changes as a result of factors such as demographic urban growth and decline, urbanization projects especially in the peripheries, shifts in the center-periphery relationship, destructions, and, finally, subsequent reconstruction, etc. (both in the present and in the past).

The recent debate in space and spatial theory in geographical as well as in social and cultural sciences has shown that we can no longer reduce space to concrete, built space such as houses, but that we should distinguish different space types, their constitutional processes, order, use, and representation. Twenty-five years ago Bernard Lepetit, a French historian, had already suggested that we should consider space as a configuration according to which a society, by ignoring or by believing to know what it does, still makes a choice among the past spatial organizations (Lepetit 1986).

As we assume that space is not a pre-set, given entity but something that is socially constructed, we propose to use the concept of "spatial relation" (instead of "space"), which may encompass the formation of spatial configurations and hierarchies arising from social interaction, but also from spatial practices and perceptions.

As this volume contains not only several case studies but also essays on methodology, it is clearly aimed at a wide-ranging readership reflecting a great variety of diverse interests. However, the one element unifying all the various contributions is focused on the theme of towns and cities. Within this historical perspective on cities, the special interest concerns the history of urban transformations with regard to the city's demographic and social aspects, but, most importantly of all, however, with regard to spatial factors. The subdiscipline, the history of urban planning, will be touched upon in cases such as when the question arises as to how certain urban forms came about. Topics concerning the perception of these forms together with practices involving the appropriation of urban spaces belong to the historical anthropological or media science disciplines. Topics concerning the possibilities of visual presentation of past and present urban spaces involve data-based cartography and geoinformatics. Taken as a whole, the papers in this volume also contribute to the problem as to how space–time changes can be adequately presented graphically and/or visually. However, this does not imply that we are concerned merely with a simple visualization or illustration of spatial or space–time phenomena, but that we are more concerned with the visualization of social phenomena, which are difficult to portray even via a textual medium. The great advantage of using data-based maps or generally computer-based methods is that social phenomena in the past or today can become visible by these methods. At the methodological level, this volume sees itself as a contribution to an epistemology of spatial change. The considerations presented here may result in the production of a new model for mapping urban changes and spatial relations referring to the past, present, and future. This new model would, therefore, be an appropriate reaction not only to the fact that the majority of the world's population will be living in towns and cities but also to the consideration that cities are social spaces as well as being both perceived and represented spaces.

Half the contributions in this volume are based on papers given at a workshop at the University of Erfurt, which took place on the 18th May, 2012 and which has provided the title for this collection of essays, involving the following authors: Bernard Gauthiez, Manel Guàrdia Bassols, Richard Rodger, Leif Scheuermann,

Ekkehard Schönherr, Wolfgang Spickermann, and Olivier Zeller.[1] Representatives from various disciplines were brought together to discuss new methods of spatial analysis and modes of representing changes in perceptions. Additionally, the volume has been supplemented by contributions given by Urška Perenič and Benjamin Vis who also participated in the conference, and by Susanne Rau, the project leader in Erfurt. Benjamin Vis's mediation enabled us to obtain contributions from Peter C. Dawson, Shawn G. Morton, Meaghan M. Peuramaki-Brown, Stephen Read, and Jeffrey D. Seibert, who were participants at the "Assembly for Comparative Urbanisation and the Material Environment" (ACUMEN)[2] and were part of the network which grew out of this assembly. These later contributions have enriched the volume by supplementing and extending these themes, yet remaining within the spirit of the original parameters set by the Erfurt conference. Even though some individual contributions were regrouped, this is the reason why we decided in favor of retaining the basically tripartite workshop structure: 1. Maps and Technology, 2. Mapping Cities: Lyons and Barcelona as Case Studies, and 3. Mapping Humanities.

The first part of this volume is mainly concerned with the technological aspects of projects already in existence which deal with the cartographical representations of cities and their development as well as with urban social space and its uses. In addition, the volume contains discourse on the methodology involved in presenting socio-spatial relationships.

Richard Rodger's presentation of the *Visualizing-Urban-Geographies* (VUG) project gives an example, which shows how historians and geographers can successfully work in collaboration with each other. This project enables the public to have online access to georeferenced historical maps and to be able to make comparisons of the maps with one another and with present-day geographies. Using Edinburgh, the Scottish capital, as his example, he explains how new historical insights can be gained by integrating additional spatial data.

Leif Scheuermann adduces theoretical considerations to the general themes on how to design and develop a co-productive spatio-temporal information system. Its central focus is on Webble technology, which can take the first step toward enabling a computer-based historical science to develop. Both the potential and the limits of the input of this technology do not seem at the moment to have been fully established.

---

[1] Georg Gartner, Professor for Cartography and Geo-Media techniques at the Vienna University of Technology and President of the International Cartographic Association—ICA, was also a conference participant. In Erfurt his paper was on the EmoMap project concerning the emotional spatial perception in navigational systems for pedestrians. Further information on EmoMap can be found in the 'Infonetz' of the Österreichische Forschungsförderungsgesellschaft: http://www2.ffg.at/verkehr/projekte.php?id=754&lang=de&browse=programm. Accessed 1st September, 2013.

[2] University of Leeds, UK, 12–13th December, 2012: "Digital methodologies for social research on processes of urban landscape development".

In the third contribution, Shawn G. Morton, Meaghan M. Peuramaki-Brown, Peter C. Dawson, and Jeffrey D. Seibert present a method to enable the visualization of socio-spatial interactions to take place in cities, which can only be investigated by archeological methods. Taking Copan, a Maya city in Honduras, as their example, they show how, despite the lack of archival material, it is possible to gain information on the spatial practices concerning the movement of actors as well as their social constitution and control.

The final contribution of the first part also arose from an archeological context. Benjamin Vis proposes the application of Boundary Line Types to investigate and categorize urban spaces which have wide variations in their social connotations and functions. In the first instance, he takes Chunchucmil, a Maya city in Mexico, as his example. However, this process can be generalized and thus be applied even to archivalized western cities such as the southern English city of Winchester, which Vis takes for his example.

With the help of studies on Lyons and Barcelona, the second part deals with the transformation of point-by-point localizable evidence (which historians have found mainly in early modern community, church, and court acts) into spatially referenced cartographical presentations of historical urban and socio-spatial questions.

From their studies lasting over several decades, Bernard Gauthiez and Olivier Zeller present their results for the early modern city of Lyons (France). The authors evaluated extensive series of municipal archives and then transferred the results into a geo-information system. As a result, the system led to a new perception of the historical city; in addition, this also meant that the spatial distribution of economic factors could now be compared with social milieus, thus providing greater clarity for the understanding of political processes.

Manel Guàrdia Bassols has compiled the research results of his investigations on medieval and early modern Barcelona (Spain) with the support of Sergi Garriga's mapping. As with the investigations of Gauthiez and Zeller, Guàrdia's results are also based on archived sources. His main interest concerns movement and stability in the city's population, but he also focuses on the influence of architectural changes on the city's general direction of development and on the relationship between social grouping and spatial differentiation.

The third part of this volume concentrates more on cultural science-based perspectives and topics. Here, the main concern is on constructions and concepts of space on the one hand and concrete urban spaces on the other, the mapping of which is connected with the specific demands of individual cases.

Susanne Rau's focus is on the subjective spatial aspects of Barcelona as experienced by travelers, who described the city in their travelogs. This also involves different techniques for capturing urban space and the various notions of its structure. Finally, the problem arises as to how far both subjectively perceived spatial relations and urban spatial transformations can be mapped.

Stephen Read's contribution takes a more philosophical stance as he is interested in the phenomenology of space and the subject-object relation in the world. Read argues against bipolar concepts, which emphasize a strict division between a subjective mind-based world on the one hand and an objective physical world on the other and in opposition to this dichotomy, he proposes an argument which allows for a reciprocal interpermeation of the two worlds. This has led to new insights into our understanding of social space and its potential for mappability.

Urška Perenič's contribution introduces a literary geographical research project entitled "The Space of Slovenian Literary Culture" to the volume. The investigation and mapping of the places of birth, life, and death of Slovenian writers has led to a broad perspective, which is more focused on interurban reference points and networks than on internal structures within cities. The result, among other things, has highlighted the spatial conditionality and dependency pertaining to the literary production of Slovenian authors.

Ekkehard Schönherr's essay is the third and final contribution on Barcelona in the present volume. He presents diverse structurizations of urban space varying according to functional requirements, which can be found in early modern sources. These results have led to some new considerations on how to map historical urban perceptions.

As so generously promised at the Erfurt workshop of May, 2012, Wolfgang Spickermann has offered his commentary on this volume. As highlighted by Spickermann, two points also seemed to be of particular importance to the editors: first, that the so-called eHumanities involve far more than the electronic evaluation of texts; second, that data-based dynamic maps have great potential for historical cultural sciences.

Finally, we are delighted to express our thanks to the Deutsche Forschungsgemeinschaft (DFG) (German Research Foundation) without whose funding of travel costs for guests from other countries, the international workshop would not have been possible. We would also like to express our gratitude to the editors of "Lecture Notes in Geoinformation and Cartography" for accepting us into their ranks, which has enabled us to present our research results to a wide readership in both a printed and in digital form. A second debt of gratitude is owed to the DFG for an extra fund to cover the publication costs. Last but not least, thanks are due to John Gledhill who, with great élan, took over the task of proofreading the English contributions and of translating individual essays. The whole project was made possible not only by an additional grant from the University of Erfurt but also because of the University's support in many other ways.

Now it is left up to the readers interested in these fields to decide how the collaboration between historical cultural sciences, cartography, and geo-information systems can be continued in the future.

Susanne Rau
Ekkehard Schönherr

Reference

Lepetit B (1986) Espace et histoire. Hommage à Fernand Braudel. *Annales ESC* 41:1187–1191

Contents

Part III Mapping Humanities

Part I
Maps and Technology

When History Meets Geography: The Visualising Urban Geographies Project

Richard Rodger

Abstract The chapter proceeds from an account of the cautious embrace of historical mapping in Britain to explain why a new emphasis on open source mapping tools provides an attractive and productive way forward for historians. More specifically, the argument is that such tools facilitate analysis of historical sources and that these can be understood and applied with a very modest investment of time while yielding new perspectives on a wide range historical data. Furthermore, since there is a historical dimension to most humanities and social science disciplines, the tools development by the Visualising Urban Geographies (VUG) project at Edinburgh University offer productivity gains for researchers in other disciplines too.

1 Structural Dimension to Historical Mapping in Britain

Historians are accustomed in their post-modern world to the 'linguistic turn' and the 'cultural turn.' As a result, the edges of their territory, and perhaps even some of their core areas, have been invaded by related humanities and social science disciplines. These friendly 'aliens', with their linguistic and cultural contributions, have re-shaped historians' intellectual space. Yet, space itself, and spatial relationships particularly, seem to present a road-block to the time-travelling historian.

Electronic Supplementary Material The online version of this chapter (doi:10.1007/978-3-319-00993-3_1) contains supplementary material, which is available to authorized users.

R. Rodger (✉)
School of History, Classics and Archaeology, University of Edinburgh,
Edinburgh, Scotland, UK
e-mail: Richard.Rodger@ed.ac.uk

S. Rau and E. Schönherr (eds.), *Mapping Spatial Relations, Their Perceptions and Dynamics*, Lecture Notes in Geoinformation and Cartography,
DOI: 10.1007/978-3-319-00993-3_1,

Why should history and geography, time and space, be in a state of tension? Arguably it might begin at school: in Britain, the national curriculum forces school pupils to choose aged 14 between geography and history as timetabling rarely, if ever, permits both. At university, geography has been colonised by geo-sciences, and humanities remains home to history. Students rarely, if ever, study both subjects and the funding algorithms for universities have favoured physical geography so that even within geography itself historical geography has been in decline in recent decades. Furthermore until about 2000 the humanities research council funding mechanisms in the British university sector favoured the 'lone scholar' rather than the research teams that featured prominently around senior professorial leadership in Germany, North America, and to some extent through the French Centre National de Recherche Scientifique (CNRS).[1] Coupled with these structural factors was the emergence in the 1960s of main-frame computer-based mapping technologies embedded in science disciplines, and the less capital intensive emergence in 1986 of DOS-based desktop versions and, in the 1990s, a Windows platform using a MapInfo package.[2] This technological dimension reinforced the 'lone scholar' emphasis—first because the heavy capital investment was not available to the humanities, and subsequently because 'distributed computing' with its desktop machines gave humanities a degree of autonomy.

Despite these developments one formidable barrier confronted the history user in addition to the not inconsiderable cost of software: the steep-learning curve of GIS was intimidatory and a powerful disincentive to the occasional user. For the social scientist using contemporary boundaries to elicit voting patterns or social inequalities, or the archaeologist or geographer with technical support, the learning curve could be flattened, or even eliminated.[3] This was not so for most historians as 'lone scholars.' As the editor of what was described as 'the first collection of case studies applying geographic information systems (GIS) to the study of history' proclaimed in 2002: 'For many scholars working in disciplines other than geography, GIS is a new and daunting tool'.[4] Nonetheless, undaunted presumably, the number of historical GIS topics had so increased that just six years later, in 2008, the same editor commented: 'The number of historians using GIS … is growing so rapidly that many of us in the field expect to see an exponential rise in GIS-based historical studies over the next decade'.[5]

---

[1] An exception was Humphrey Southall's large scale 'Vision of Britain' project. See http://www.visionofbritain.org.uk/footer/about.jsp.

[2] See for example, (Foresman 1997).

[3] Gregory (2002), online at hds.essex.ac.uk/g2gp/gis/index.asp where a number links provide historical case studies that can be viewed.

[4] (Knowles 2002), xi.

[5] (Knowles 2008), 2.

2 Maps in Minutes

Historical mapping has so far relied almost exclusively on GIS. In turn this has required digitised and geo-referenced maps, and databases designed to retrieve information on selected parameters. 'Polygons (representing the exact shape and location of administrative units) and attributes associated with those polygons'[6] were what underpinned historical GIS and presented a barrier to the general user. The Visualising Urban Geographies (VUG) approach (see http://geo.nls.uk/urbhist/) developed at Edinburgh University reasoned as follows: if an accessible tool was available to map historical data it would transform the research and teaching potential of many individuals who are currently deterred by the steep investment required to use GIS. So to liberate the spatially challenged from these 'daunting' GIS tasks—whether students, lecturers, or the general public—the VUG project developed mapping tools that could be used in conjunction with some of the 20,000 digitised historical maps held by the National Library of Scotland covering the period 1560–1928.[7] Maps created in this way and superimposed on a historical map chosen by the user would be developed in a matter of minutes. Technically, the 'mash-up' of open source web-based application programming interfaces (APIs) operated in the background and, rather like programs used every day, historians only needed to be familiar with simple cut and paste routines (Fig. 1).

Historians deal with data in two main forms: firstly, and most often, they rely on addresses since tax, employment, births, marriage and death, financial transactions such as mortgages and credit rating, and many other everyday practices are linked to place of residence. In past times, however, lists of commercial debtors, types of employment, and locations of places of worship, have been stock in trade for historical analysis, but inspection of lists is laborious and the spatial dimension is difficult to identify unless the historian is very familiar with the terrain. The second main form of historical data is compiled in districts—polygons formed by church, state, municipal, political, ethnic, and other bounded regions to record the frequency of disease, deaths, electoral characteristics, migration and the character of an area or neighbourhood. Both types rely heavily on spatial elements, but are difficult to manipulate where the data set is voluminous.

So since much of the address-based data is collected from archival records, spreadsheets form a core element of VUG tools. Cutting and pasting addresses, together with columns for other data (notes, pictures, archival source references) into a Yahoo geo-coder results in the generation of further fields for latitude and

[6] (Gregory and Southall 2002), 120.

[7] Not all of these maps are geo-referenced.

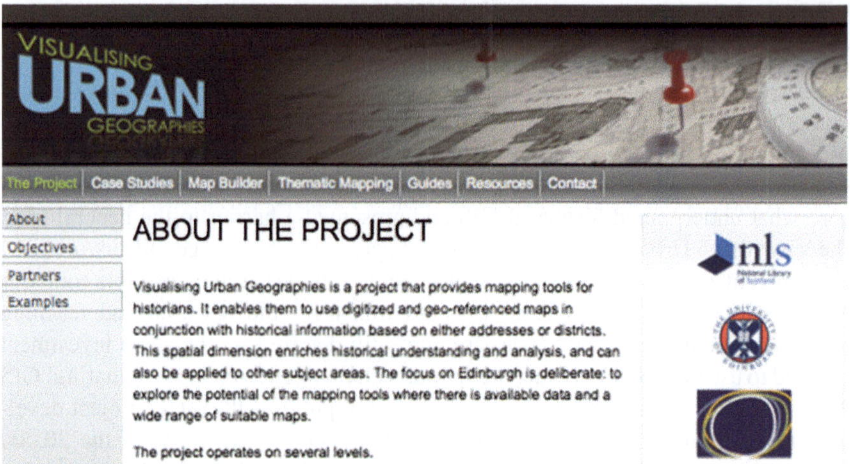

Fig. 1 Visualising urban geographies homepage

longitude. These locations can then be viewed in Google maps, or saved as a kml file to generate plots in Google Earth, or used in conjunction with geo-referenced historical maps of the period.[8]

More specifically, Fig. 3 shows the distribution of Edinburgh solicitors in 1861 and 1911, and this shows a drift of solicitors towards the main city centre streets during this time. This raises questions about the nature of legal firms, the complexity of the law over time, and a range of related research questions that would be difficult to identify otherwise. For the historian an inspection of addresses in trade directories would not itself reveal such a pattern so the mapping exercise, achieved in about 20 minutes from start to finish improves research productivity considerably.

The geo-coding routine also allows users to calculate the distances of each observation from a selected point, and this provides a measure of dispersion in the data to be established. For example, are retailers such as tobacconists more concentrated than opticians? We may speak of a 'jewellery quarter' in Birmingham but are other activities just as concentrated—or just less obvious? Measures of this kind enable the historian to develop hypotheses about the spatial distribution of social, economic, political and cultural activities that relate directly to archival sources. Users have found these tools helpful in understanding the concentrations of Liverpool moneylenders in the inter-war period and of Dundee prostitutes

[8] Batch geo-coding is not without difficulties. Like most datasets, standardisation using street, city and country column headings to ensures that Edinburgh is considered to be in Scotland, not the USA. Where streets have been obliterated, it is possible to develop 'reverse geo-coding'—where the location in the city, if not the historic street, can be identified, so that a latitude and longitude or a post-code can be assigned. This is a laborious process where there are many such locations.

during the same period. In a more extended study, membership records of several clubs and societies in Leicester since 1950 have been analysed to test the hypothesis that suburbanisation in Britain diminished the coherence of civil society.[9] Another study considered the introduction of gas lighting to Edinburgh in the 1820s, posing the question: was this more prevalent in affluent areas where it was affordable or in poor areas of the city where policing and urban management were the priority? It is worth stressing again that the use of a historical base map with its physical features and buildings of the period provides a more appropriate context for the data than a present day Google map.

The VUG project brings other analytical tools to the assistance of the historian. Precise metric measures of distance, area, altitude and gradient are features that add considerably to the interpretation of contemporary topography and scale. It is particularly easy, when using two-dimensional maps, to overlook the combined significance of distance and gradient in earlier periods, and the sheer manual effort involved in many occupations and daily activities. Lifting and carrying over considerable distances and heights fundamentally influenced human actions and interactions in ways that our contemporary preoccupation with private cars and public transport systems largely discount. Importantly, too, a focus on the 'walking city' provides a nuanced understanding of how individuals went about their business, the routes they used to avoid scrutiny or congestion, and served the users' purposes. No wonder, then, that in the West Bow leading out of the Grassmarket to the Lawnmarket near Edinburgh Castle, a distance of 160 m was associated with a 12 m incline (a 14 % gradient) for carts and carters and presented such serious conditions underfoot in winter especially that an improved route was funded by the City Council to reduce the risk of accidents and the spillage of goods on to the cobbled streets.

In another case the decision to build housing on a hillside had much to do with the gradient since the costs increased disproportionately with steeper inclines and this defined profitability. The VUG mapping tools, particularly those associated with gradients, improve an understanding of such business decisions and do so on geo-referenced historical maps that provide precise topographical details.

3 Map Layers

The VUG tools provide a powerful analytical tool for historians. However, the productivity of the tools is enhanced by deploying them in conjunction with historical maps of a time period appropriate to the data. In most European towns and cities map-making in the nineteenth century was itself a flourishing business. Municipal authorities, large and small, sought to understand and manage their jurisdictional areas. These maps, commissioned by the city and the state, are

[9] See for example, Balderstone (forthcoming).

increasingly available in digitised and geo-referenced form, and so historians can deploy their data in conjunction with maps that correspond closely to the era to which they relate.[10] In the VUG project 25 geo-referenced maps covering the period 1765–1940 have been made freely available as downloadable JPEG images, KML SuperOverlays, Tile Map Services and Web Map Services.[11] These include the most detailed and accurate maps of the city based on a direct survey by William Edgar (1765), John Ainslie (1804), Robert and James Kirkwood (1817, 1821), John Wood (1831), James Kay (1836), Bartholomew's (1865, 1882, 1891, 1893–1894, 1902, 1907, 1912, 1917–1918, 1919, 1932–1933, 1939–1940), W. and A. K. Johnston (1905–1906, 1910–1911), and two Ordnance Survey series maps (1853–1853, 1877).[12]

The VUG project is organised around the concept of a Map Builder. This is a customised open source mashup framework for working with maps and data. That is to say: the user's data forms the basis of a layer superimposed upon a historical map chosen by the researcher to match as closely as possible the date of the data. Examples were presented in Figs. 2, 3, 4, 5 when customised VUG Map Builder tools were used to plot addresses, and to identify distance, gradient, area and altitude in relation to a map of the period.

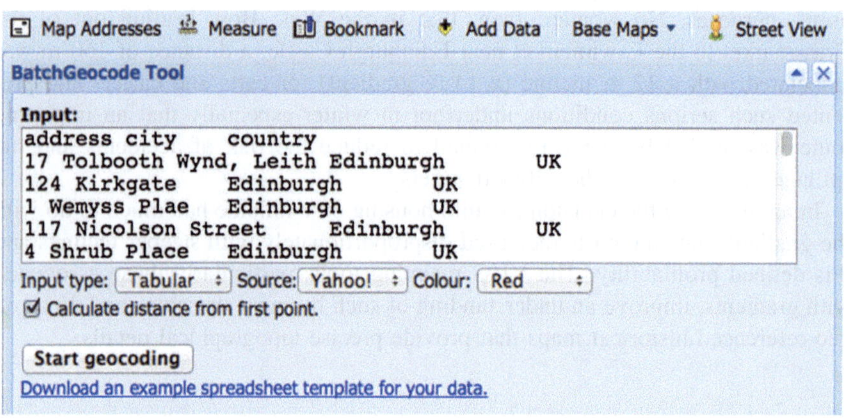

Fig. 2 Batch geo-coding addresses near here; see also mp4 geocoding

[10] If not already undertaken by an archive or other repository, the VUG project website provides a drop-down 'Guides' menu with step-by-step help on how to geo-reference maps. See http://geo.nls.uk/urbhist/guides_georeferencing.html This can be done using ArcGIS, QuantumGIS, and Georeferencer methods for cropping, choosing co-ordinates, adding control points and transformations. For further methodological considerations see Ballett (2006), Boutoura and Livieratos (2006).

[11] These maps and WMS layers are all based on the British National Grid, OSGB 1936 (EPSG:27700) coordinate system.

[12] http://geo.nls.uk/urbhist/resources_maps.html

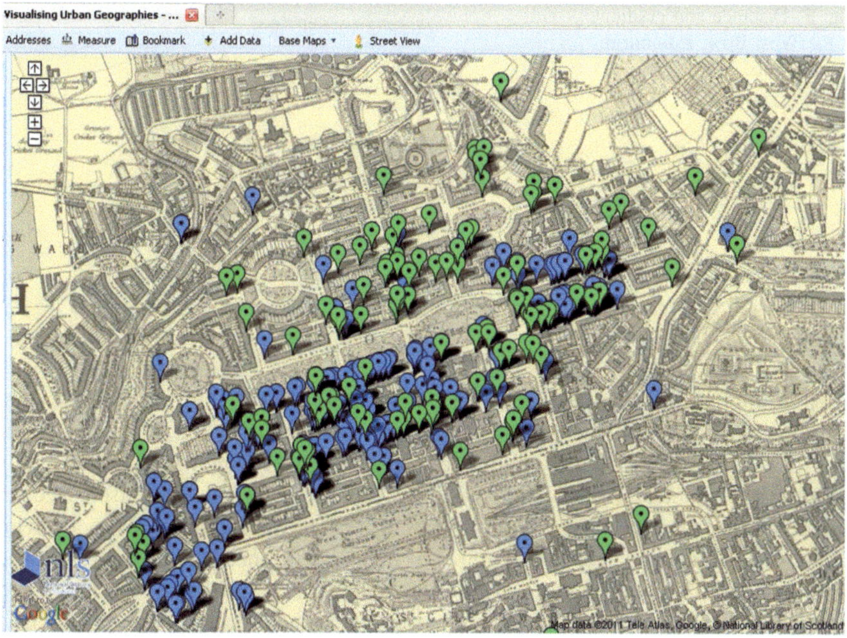

Fig. 3 Edinburgh Solicitors 1861 (*green*) and 1911 (*blue*) (Color figure online)

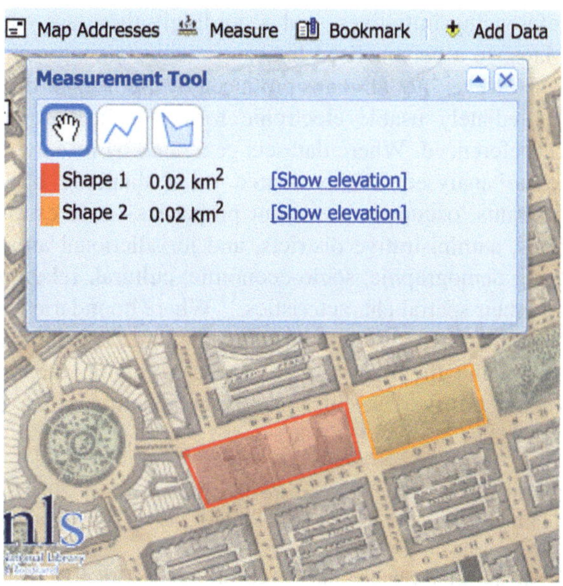

Fig. 4 Distance and area measuring tools—see also mp4

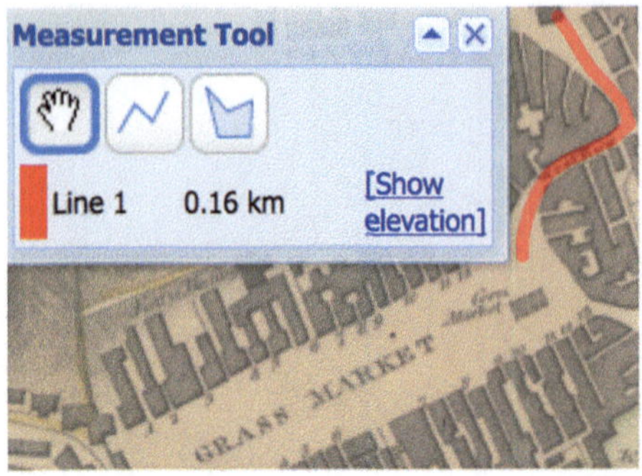

Fig. 5 Distance and gradient in the West Bow, Edinburgh before 1830

How, though, are the geo-referenced maps uploaded in the first place so that they can appear as choices in the menu bar, as shown in Fig. 6 Map Builder provides the basis for this; it requires a Google account which means that this is where users' data is stored—not on the VUG server! First time users see an instruction to 'Add Map' from the VUG Map Builder menu; this leads to a screen with an empty spreadsheet that requires a map to be named and an address to be provided, and finally leads to an instruction 'Publish as a web page' for the named map. Such published links are, of course, portable as with any web link.[13] This sequence enables users to add further maps to correspond with their historical data. The result is that geo-referenced historical maps are customised so as to give a suite of options that correspond closely to the researcher's historical data set. The National Library of Scotland has been at the forefront of such digitising and geo-referencing initiatives, and accordingly there are many existing options to 'Add Maps' to a personalised list.

However, not all maps produced by archives or data libraries are available in an immediately usable electronic form. Physical copies need to be scanned and georeferenced. Where datasets generated by researchers from archival sources are to be analysed and presented on a thematic basis—debts, prison sentences, migrants, occupations, vacant properties—it is essential to capture local boundaries, administrative districts, and jurisdictional areas that will reveal the underlying demographic, socio-economic, cultural, religious and political relationships and their spatial characteristics.[14] Where boundaries are subject to change over the long historical sweep, as is often the case, it is necessary to record these changes by tracing them on a map and then saving the boundary lines as a separate vectorised overlay. Such newly created maps can then be uploaded using the

[13] This sequence of steps is provided in detail and with several screen shots as part of the information on customising Map Builder and on how to publish maps as web pages See http://urbhist.nls.uk/mapbuilder/.

[14] In the 21st century the Office of National Statistics and the Scottish Government use respectively over 31,000 and over 6,500 geographical units to map various social characteristics and overall indicators of multiple deprivation. See http://www.scotland.gov.uk/Resource/0041/00410727.pdf.

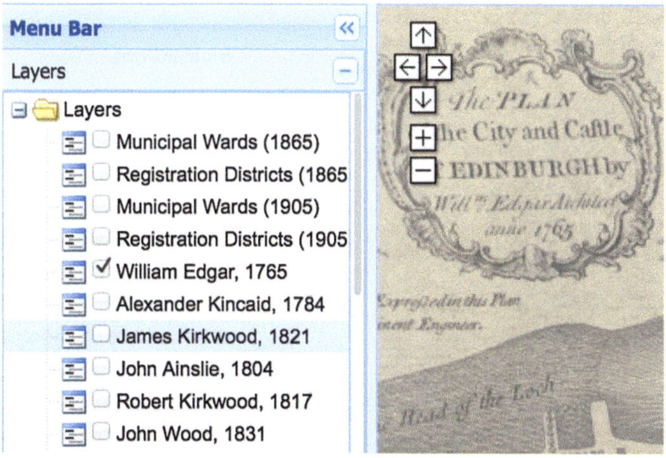

Fig. 6 Map layers

'Add Map' option and will appear in the VUG Map Builder sidebar as options. With administrative areas available then spatial analysis can be undertaken using 'choropleths'—shaded polygons that capture the graduated nature of the characteristics.

In Edinburgh a number of such administrative boundary polygons have been captured by the VUG project and made available as shape files and kml files. These templates include the changing registration districts, wards and municipal boundaries for Edinburgh from 1865 to 1912, and the various extensions of the city limits from 1685 to 1885. In addition, an exemplar of other types of boundaries, Kirkwood's detailed cadastral map of landownership in Edinburgh in 1817 has been captured in a database that also contains landowners' names. The area of each plot is also generated as a by-product of geo-referencing. Similarly, a range of socio-economic data has been recorded and made available, and each item of data provides background content to the dynamic thematic maps in the Map Builder application. (See Fig. 7). Population details from the censuses from 1861 to 1921 have been recorded, including numbers of males and females; inhabited houses by registration district for each census and house rental values by sanitary districts have also been recorded for 1874 (see Fig. 8). In addition, details of the occupants and their addresses in the selected Edinburgh housing settlements in 1891 have been made available. Overall, 25 maps were geo-referenced, 23 administrative boundaries were vectorised, and 16 socio-economic datasets provided as part of the VUG project. In each case the intention is to demonstrate how boundary based maps, used in conjunction with historical data derived from archival sources, enriches historical understanding and interpretation.

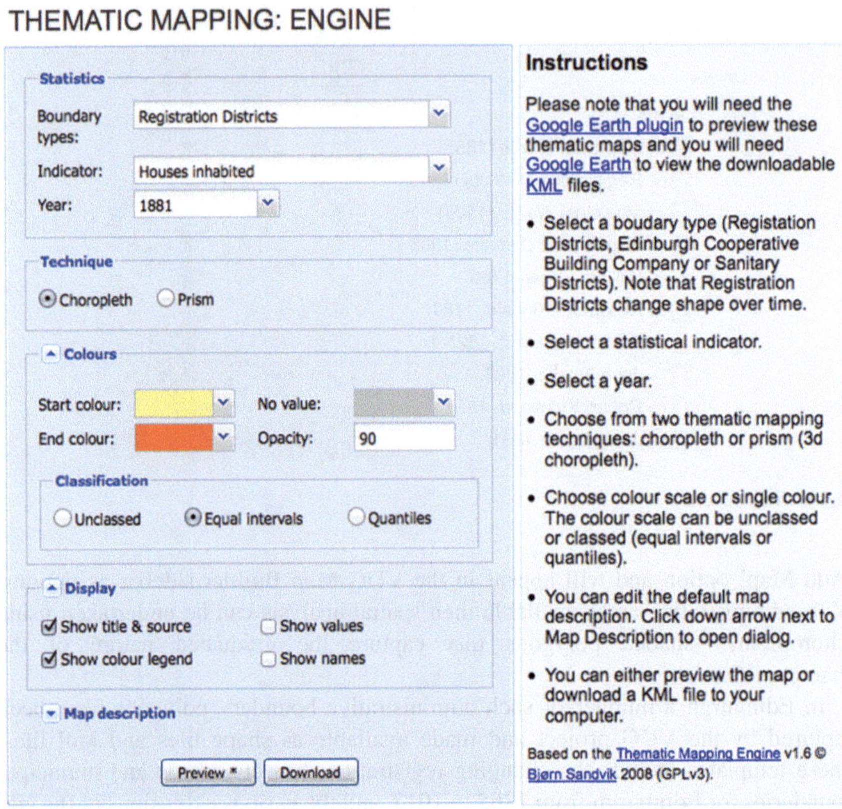

Fig. 7 Map builder and the thematic engine

4 Dissemination Strategies

The programme of disseminating project results differed in its intensity and extent from that originally planned. Whereas an important element in the original plan was to take the project to a number of archives and libraries, it transpired that a number of public events were hosted by other organisations so it was possible to make contact directly with large numbers of delegates and members of the public. This was the case, for example, at road shows within Scotland (Perth, Inverness, Dunfermline), through public events hosted in Edinburgh, and webcasts and academic conferences.[15] An active follow-up programme of leafleting reached the memberships of organisations and schools, and regular features in *Cairt*, the Newsletter of the Scottish Maps Forum, engaged with the geographical and

[15] See for example, VUG Workshop (6 Dec 2010) http://geo.nls.uk/urbhist/workshop.html; VUG Launch (24 Feb 2011) http://geo.nls.uk/urbhist/launch.html.

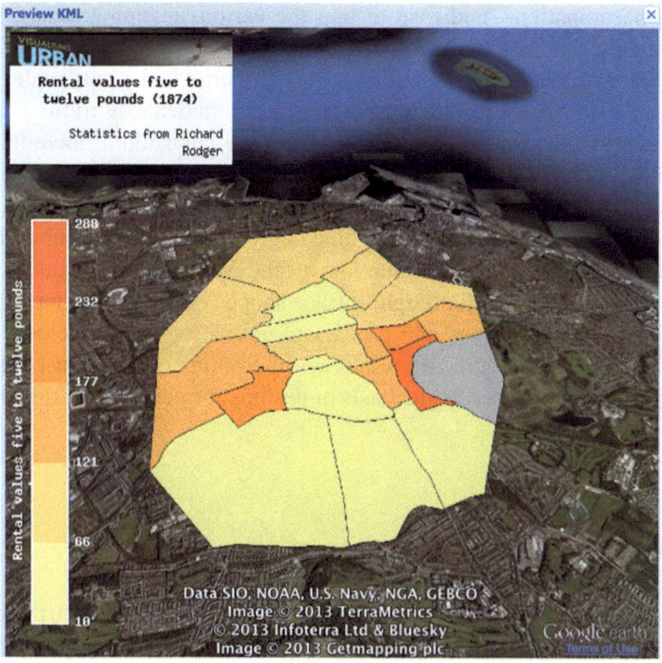

Fig. 8 Distribution of house rentals (£5–12) by Edinburgh districts, 1874

cartographic interests of the public. Dissemination to the scholarly community involved presentations to national and international meetings largely as planned. An early decision to present a paper at an International Digital Cartographic conference in Vienna enabled the technical officer to make and, subsequently, develop important personal contacts.

The lists of contacts and dissemination activities far exceeded those anticipated. Expertise acquired in developing the Map Builder tool and the developing NLS list of georeferenced maps proved popular and widely appreciated, and technical dialogues arising out of the VUG project developed within and between a variety of organisations and projects, including Information Services personnel at Edinburgh University, EDINA, and several digitisation projects in European Union countries. Disciplines outside the humanities also showed an interest in using the mapping tools, specifically in medicine, psychiatry and criminology. A contribution to the Digital History seminar at the Institute of Historical Research provided an opportunity by means of a webcast to disseminate the VUG outputs through Twitter, and other social networks reflected this wider reach. Google Analytic data shows over 70,000 pageviews and almost 50,000 unique page views in a three-year period between March 2010 and March 2013. The Guides and Help Notes appeared to be particularly appreciated.

Student participation—and road testing the various mapping tools—was originally planned as a minor part of the VUG project but the concept and utility of mapping historical data soon caught on with Edinburgh University undergraduates following a 'History in Practice' seminar with 20 students. Various requests for assistance with student projects at all levels of the curriculum, including several PhD candidates, then developed from different universities.

An unexpected project outcome resulted from a meeting with an undergraduate from Edinburgh College of Art. In the final degree show at the Edinburgh College of Art, and arising out of a presentation by the VUG team, a student has used historical maps of central Edinburgh to trace shapes on glass, and backlight these to provide illuminated shapes and profiles of courts, wynds and closes. Another creative element involved tracing relatives' migrations over the last century and using the patterns developed as the basis of designs for pen and ink drawings. This creative work is ongoing.

5 Postscript

Arising out of the energy generated by, and public interest in, the VUG project, a group of 30 contributors and 5 partners[16] has successfully proposed a three-year funded project to the UK Arts and Humanities Research Council.[17] 'Mapping Edinburgh's Social History (MESH): A Capital Digital Resource' will develop further the tools associated with the VUG project and disseminate these by means of a digital atlas that will provide important guidelines and protocols for similar publications and websites for other towns and cities. Together with digital data and maps the intention is to extend and enrich the public's local historical interests and participation. A fully developed scholarly agenda also forms part of the MESH project.

In our modern world, post-codes define the activities of mail order firms and fleets of commercial delivery vans, and sat-nav equipment enables travellers to hurtle through the countryside oblivious as to how places relate to one another. Historians, and humanities researchers generally, are increasingly able to refine a spatial 'turn' or perspective using data which has sometimes remained obscure. Digital mapping now renders this amenable to systematic analysis.

Acknowledgments The 'Visualising Urban Geographies' project was funded by the Arts and Humanities Research Council, UK, under the Knowledge Transfer Fellowship Scheme, Grant AH/G017077/1. See also Rodger et al. (2010).

[16] The partners are the National Library of Scotland, World Heritage Centre, Edinburgh City Council, Royal Commission on Ancient and Historical Monuments in Scotland, and a private firm of conservation architects.

[17] AHRC AH/K002457/1.

References

Balderstone L (forthcoming) Semi-detached Leicester: social and cultural connections in suburban Leicester. In: Rodger R, Madgin R (eds) Leicester: a modern history. Lancaster

Ballett C (2006) Georeference in the analysis of the geometric content of early maps. e–Perimetron 1(1):32–42

Boutoura C, Livieratos E (2006) Some fundamentals for the study of the geometry of early maps by comparative methods. e–Perimetron 1(1):60–70

Foresman T (1997) The history of GIS (Geographic Information Systems): perspectives from the pioneers. Harlow

Gregory I (2002) A place in history: a guide to using GIS in historical research. Oxford. Online at hds.essex.ac.uk/g2gp/gis/index.asp

Gregory IN, Southall HR (2002) Mapping British population history. In: Knowles AK (ed) Past time past place

Knowles AK (ed) (2002) Past time, past place: GIS for history. ESRI, Redlands

Knowles AK (ed) (2008) Placing history: how maps spatial data and gis are changing historical scholarship. ESRI Press, Redlands

Rodger R, Fleet C, Nicol S (2010) Visualising urban geographies. e-Perimetron: Int Web J Sci Technol 5(3):118–131

Websites

http://geo.nls.uk/urbhist/guides_georeferencing.html (Step-by-step help on how to geo-reference maps)

http://geo.nls.uk/urbhist/launch.html [VUG Launch (24 Feb 2011)]

http://geo.nls.uk/urbhist/resources_maps.html (Maps used in VUG)

http://geo.nls.uk/urbhist/workshop.html [VUG Workshop (6 Dec 2010)]

http://urbhist.nls.uk/mapbuilder/ (Map Builder and on how to publish maps as web pages)

http://www.visionofbritain.org.uk/footer/about.jsp (Vision of Britain)

http://www.scotland.gov.uk/Resource/0041/00410727.pdf [Summary of key findings from the Scottish Index of Multiple Deprivation (SIMD 2012) published on 18 Dec 2012]

Thoughts on a Web Based Co-productive Spatio-Temporal Information System

Leif Scheuermann

Abstract With the new technologies—summarized under the buzzword "Web 2.0"—the options of web based collaborations have increased strongly. In this context large projects on a voluntary basis like Open Street maps or Wikipedia have formed platforms to accumulate enormous quantities of data. Also in e.Humanities there are trends and plans towards such crowd sourced information systems and, like their big brothers, they face some basic problems: first how to bring together existing but very diverse distributed data in one system and second, how to guarantee the quality of the data—respectively, how to prevent the misuse of a system. The following paper provides some basic ideas for this discussion, without having the claim of being unique or in any way complete. The first part will cover the topic of how to implement a co-productive historical spatio temporal information system using webble technology, and the second part focuses on the question of quality management in such a system.

1 Introduction/Requirements

Since the introduction of the term "Web 2.0" by di Nucci in 1999, the idea of the collaboration of large groups of users has determined the development of web based applications. Today social media platforms such as "Twitter" or "Facebook" are as omnipresent as the different Google services like "Google maps" or "Google earth". Therefore, it is no wonder that historians (professionals as well as laymen) are trying to jump on the bandwagon, either by using the big platforms or by creating their own. In this dynamic, the question of what is already existing and actually needed is often neglected. During the last few decades numberless

L. Scheuermann (✉)
Interdisciplinary Center of E-Humanitites in History and Social Sciences, Max Weber Center for Advanced Cultural and Social Studies, University of Erfurt, Erfurt, Germany
e-mail: leif.scheuermann@uni-erfurt.de

S. Rau and E. Schönherr (eds.), *Mapping Spatial Relations, Their Perceptions and Dynamics*, Lecture Notes in Geoinformation and Cartography,
DOI: 10.1007/978-3-319-00993-3_2, © Springer International Publishing Switzerland 2014

projects digitized huge amounts of data and often provided them via the internet. A central challenge for the future will be to combine these insular sources and to enable users to add their own sources as well as to ask their own questions to them. The first requirement on a comprehensive historical information system is to provide interfaces for all kinds of distributed data as well as tools for joint analysis and visualization.

Besides this more technical challenge, a major problem of these new co-productive information systems is the quality management of the data. Especially in using big data, the researchers have to rely on the accuracy of the sources. There have to be strategies to ensure the quality and to document the origin of the data. Furthermore, the objective of the data acquisition and the data model has to be apparent to ensure a correct understanding of the data. Thus a set of metadata and a metadata model has to be provided to enable the user to handle the co-productive historical spatio-temporal-information system properly.

Due to these now specified requirements, two topics shall be discussed below: on the technical side, webble technology as a middleware that is able to combine different distributed tools and data and on the methodical side, metadata and the Conceptual Reference Model of the International Committee for Documentation of the International Council of Museums (ICOM—CIDOC CRM).

2 Webble Technology

One of the main features of the "Web 2.0" technology is the combination of different web based applications in so called "mashups". Via web application programming interfaces (API), distributed media can be (re-)combined to create new content. Usually these APIs are libraries containing specifications for the data structures, methods, classes and the routines strictly defining the coverage of the interchange between the applications, which limits the compatibility. The exchange itself is only content based and the applications such as "Google Maps" etc. remain unaffected. So users can integrate their data, perform some analyses provided by the basic application and its APIs, and display them.

Here the meme-media approach[1] takes a step further, by wrapping the applications with reactive media-component representations, which act as interfaces. With this object oriented approach not only the content, but also the whole application is integrated into the meme media framework; therefore every user can adapt it for his/her own purpose and extend it with new functionality. At the same time most diverse and widely distributed wrapped data can be included without changing the original. The web based form of the meme-media approach—webble-technology (web based lifelike entities)[2]—allows the combination of wrapped objects—called

[1] A basic introduction to meme-media: Tanaka (2003).

[2] Basic introductions to webbles: Arnold et al. (2013), Tanaka and Kuwahara (2012).

webbles—simply via drag and drop. These combinations create new compound webbles, which can be combined with others and so on. The communication between objects is controlled by slots, hierarchically ordered, configurable bilateral stream channels, which can transport any types of data from one to one or more objects. Via these slots the receiving webbles can be directly manipulated by the sender. So the change of one webble influences all the connected webbles. To give a very basic example: the size of one webble with the shape of a rectangle can be determined through the incoming data from a second webble, which is a slide control.

2.1 First Ideas on a Webble Based Spatio-Temporal Information System

This on first sight fairly simple working principle can now be used for the construction of highly complex compound structures like a spatio-temporal information system. But what would such a system look like? First of all, there has to be some input. To bring them into the system, input webbles are needed. Basically it doesn't matter if the data comes from local sources, from the internet, or is directly typed into a text field and it is of no relevance how complex its structures are, as long as the webble embeds it in the right slots. Next the framework needs visualization tools. For a spatio-temporal information system this means some kind of a map and a timeline, but also text fields, tables, charts and graphical representations such as network diagrams are conceivable. Because of the openness of the system, the users can even include their own custom-made visualization webble or build a compound webble from the existing ones. An interesting feature about webbles is that they are not only for the mere display of the data, but also can be used as selection tools. For example, the map can be used for the geographical, or the timeline for the temporal localization. This manipulation of one of the visualization webbles can then have a direct effect on all the others and even on the input webbles, which form new requests to the data sources. Thus the boundaries between visualization and analysis tools become permeable. But of course there should also be specific analysis tools such as spatial analysis tools and agent-based simulation models.

3 Quality Management

After this brief outline of a co-productive spatio-temporal information system, the following paragraphs will broach the issue of the quality management of the data used in the system. This topic is one of the most important problems especially for co-productive voluntary based projects, but also for frameworks using various distributed data, as, for example, when embedding web pages. At the first sight, the large public based projects seem to be self-organized and the data public property. At a second glance the self-organization is gradual and the data is always related to actors, even if these are not necessarily related to an individual. The responsibilities for the quality of the data therefore lie in three hands:

- The creator of the data, who has not only to provide the data, but also to document their origin and structure and to classify them.
- The user, who is responsible to inform himself about the origin of the data, decides to trust the data and has the possibility to include, exclude and combine data according to his personal view/choice.
- The framework of the project, with its rules and structures that coordinates the asynchronous communication.

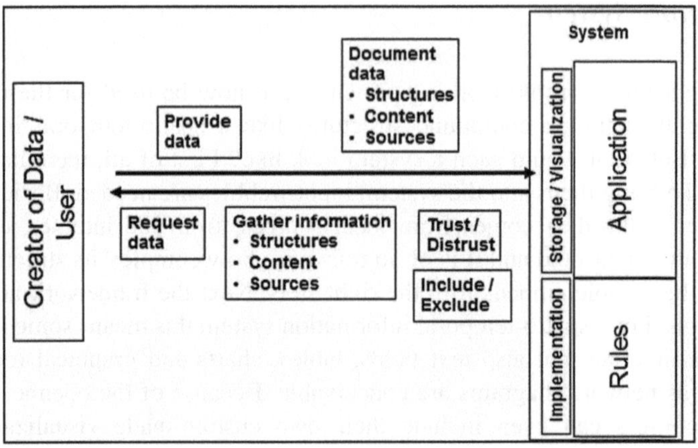

In the end, it's the duty of the system to provide a framework for a responsible research process. It is not a question of whether or not self-organized collections of geographic data are still maps, but of how a framework can ensure this and how the individual responsibilities of creators as well as users can be warranted. Therefore, as the basis of a web based co-productive spatio-temporal framework, two particular topics have to be discussed more intensely: meta data and data models.

3.1 Meta Data

To guarantee quality standards, there has to be a catalog of supplemental information for each data set. This metadata has to be visible to the users and enable each user to trust or distrust the data set.

To be able to understand the data, a brief description of the objects, their classification and the exact circumstances under which they were collected, is needed. The reason for this is the often neglected fact, that every digitalization project has a basic purpose and aim, which determines its design. Not knowing this purpose will lead to a misinterpretation. For example, the data for one of the big databases on Roman epigraphy—the "ubi erat lupa"[3]—was mostly provided by museums. Hence all the objects preserved in the museums are included—even those not yet published elsewhere. On the other hand, because of the nature of the project, some only literarily documented monuments are not included. Mapping this data gives us an image of the spatial distribution of the inventories of the museums, but not necessarily the distribution of all the monuments found or of the Roman settlements in an area. This is explicitly not the fault of the database, but a restriction by its scope, which has to be represented in the metadata.

Next, every data set needs legal notes containing the producer of the data and the person in charge, a contact, the terms of a license for the use of the data (e.g., Creative Commons Licenses) and a list of all the sources that were used for the creation of the data sets. It must be clear, especially for the use of distributed web based data, as to who owns them and who is responsible for their validity in order to respect the copyright.

And of course there has to be a temporal and spatial localization, which provides the basic values for the requests as well as the accuracy of the data. The metadata has to include the notation of the geographic coordinate system, the encoding, the data format and a list of all attributes with a separate description and data types.

3.2 Metadata Models and Ontologies

Last but not least, one of the main problems for a spatio-temporal information system using distributed data sources lies in the different data models describing the databases. As mentioned before, if the topic and basic structures of a database are not clarified and expressed in form of metadata, there can be neither an exchange of data nor can the quality of the data be evaluated. But even with a proper documented structure, the system itself can't handle the different data structures. So the question is not only how can data be described properly but how can this description itself be understood by the framework beyond the borders of language and technologies? The answer to this is to be found in so called

[3] http://www.ubi-erat-lupa.org. Accessed 30 July 2013.

ontologies—meta data models containing rules of integrity, a specific standardized shared vocabulary and rules as to how these elements are related to each other.[4] Each ontology has a scope or domain of interest and is itself related to others via higher ranked ontologies. For cultural heritage information—mental as well as material—the most used domain ontology is CIDOC CRM (International Committee for Documentation Conceptual Reference Model)[5]—created and provided by ICOM (International Council of Museums). Since 9/12/2006 it has been official standard ISO 21127:2006 and currently includes 90 classes of entities and 152 properties[6] describing all aspects of a cultural heritage. In a first step, the data structure of the sources must be understood and formalized with the entities and their relationships. The result will be a diagram imaging the data structure of the source, which is to be encoded, for example, in a XML file or a RDF schema and connected to the data sets. Now the information of the sources is compatible with any other sources using the same ontology and the framework in which they are embedded has now the possibility to correlate them to each other properly. Concurrently the data of the sources remains untouched in its structures and form. It can even be left physically at its server of origin, as long as a wrapper combines the embedded data with the descriptive files.

4 Conclusion

In the previous pages a few ideas on how to design and develop a co-productive spatio-temporal information system have been presented. The text does not make any claim to be complete or comprehensive, but wants to broach the issue of some important requirements and possible solutions. On a methodological side, this is the question of quality management and compatibility of distributed data stored in different environments and on a more technical side, the design and composition of a webble based spatio-temporal information system.

The implementation of such a system is still to be done and the next few years will show in which frame it can be realized. The necessity to get away from the traditional monolithic insular applications—local or web based—is unmistakable. The next step in the development of a computer based historical science is to bring the amount of digitized sources which has already become immense and the various analysis tools together in a dynamic and open manner. The presented ideas on a spatio-temporal framework might be helpful on that way.

[4] A more detailed introduction to ontologies in general and CIDOC CRM with further literature: Scheuermann (2006).

[5] http://www.cidoc-crm.org/. Accessed 30 July 2013.

[6] CIDOC CRM Special Interest Group (2013).

References

Arnold O, Spickermann W, Spyratos N, Tanaka Y (2013) Webble technology. In: Proceedings of first webble summit, WWS 2013, Erfurt, Germany, 3–5 June, 2013. Springer, Berlin (Communications in Computer and Information Science 372)

CIDOC CRM Special Interest Group (2013) Definition of the CIDOC Conceptual Reference Model Version 5.1 (May 2013). Downloadable at: http://www.cidoc-crm.org/docs/cidoc_crm_version_5.1-draft-2013May.pdf. Accessed 30 July 2013

DiNucci D (1999) Fragmented future. Print 53(4):32

Scheuermann L (2006) Ontologien in den historischen Wissenschaften. Hist Soc Res 32:308–316

Tanaka, Y (2003) Meme media market architectures: knowledge media for editing, distributing and managing intellectual resources. IEEE Press, Piscataway, Hoboken

Tanaka Y, Kuwahara MN (2012) Webble world user's and developer's guide. Downloadable at: http://cow.meme.hokudai.ac.jp/docs/WebbleWorldManual.pdf. Accessed 30 July 2013

Peopling the Past: Interpreting Models for Pedestrian Movement in Ancient Civic-Ceremonial Centres

Shawn G. Morton, Meaghan M. Peuramaki-Brown, Peter C. Dawson and Jeffrey D. Seibert

Abstract The study of complex societies, in particular urban spaces such as those of the ancient Maya, can effectively focus on the human interactions and entanglements that animated such locales. Further, many of the concerns related to crowd dispersal, pedestrian traffic patterns, the constitution of community, and socio-spatial control that underlay spatial analyses of modern urban centres were equally valid in past, non-Western, urban centres. From space syntax to agent simulation and crowd modelling, this paper adopts a methodological 'train of thought' with origins well outside the archaeological mainstream that may be applied in the creation of explanatory/exploratory models for socio-spatial interaction. Within Maya studies (and indeed, other ancient contexts), these models may be profitably invoked to direct research toward a deeper understanding of how the ancient Maya may have actually lived within the monumental built environments that so strongly define them in both popular and professional consciousness. The unit of analysis in all such approaches is the plano-metric representation of architecture and space. In concert with the other papers presented in this volume, particular attention is focused on the analytical consequences (both opportunities and limitations) of such mapping. The Classic Period centre of Copan, Honduras, has been adopted as a case study.

S. G. Morton (✉)
Department of Archaeology, University of Calgary, Calgary, Canada
e-mail: sgmorton@ucalgary.ca

M. M. Peuramaki-Brown
Department of Anthropology, McMaster University, Hamilton, Canada
e-mail: pbrownm@mcmaster.ca

J. D. Seibert
Cataraqui Archaeological Research Foundation, Kingston, Canada
e-mail: jseibert@carf.info

P. C. Dawson
Department of Archaeology, University of Calgary, Calgary, Canada
e-mail: pcdawson@ucalgary.ca

S. Rau and E. Schönherr (eds.), *Mapping Spatial Relations, Their Perceptions and Dynamics*, Lecture Notes in Geoinformation and Cartography,
DOI: 10.1007/978-3-319-00993-3_3,

1 Introduction

> Every city has a sex and an age… Rome is feminine. So is Odessa. London is a teenager, an urchin, and, in this, hasn't changed since the time of Dickens. Paris, I believe, is a man in his twenties in love with an older woman… She also is Paris, and if every city has its own unique smile, in Paris it is hers.
> John Berger (1992: 101–102).

Cities are alive. For the poet, there can be no question of this fact: being born, growing, changing, and maturing, each city has a life and character all its own. Recent, exciting work by Luís Bettencourt and Geoffrey West (Bettencourt et al. 2007, 2009; Bettencourt and West 2010; Kühnert et al. 2006) has added scientific rigour to the poet's observations by demonstrating that biological scaling laws also apply to cities. Notably, from poet to physicist, it is the people who are the lifeblood of the city; thus, it is their movements and their interactions that provide the pulse.

These essential characteristics (i.e. the movement and interactions of people) have long been absent in the abandoned cities of the ancient world. In archaeology, cities are dead; therefore, there has been a tendency to treat cities as static entities. Diachronic studies that emphasize the historic development of structures or groups of structures are common, as are studies that emphasize the ideological principles referenced in architectural design and the technological innovations and limitations that may have underlain the formal characteristics of structures and spaces. Unfortunately, all of these studies remain apt to produce views of the built environment in which the day-to-day entanglements of people, places, and things that animate such spaces, along with the related principles and concerns of urban design, are rarely contemplated (Peuramaki-Brown 2012). The cityscape as a whole is rarely woven together through the movement of actors as a dynamic, functioning, living, urban space.

This volume comments on the proclivity to treat the cityscape as a static set piece, which has been a proliferating tendency in Western social sciences since at least the 15th century (Scott 1998). It seeks to challenge this position by explicitly exploring the topic of complexity within the dynamic socio-spatial cityscape together with its possible methods of representation. As showcased by many of the papers in this volume, the authors borrow from (or adapt) methods and theories developed for understanding modern urban spaces, and apply these to past urban environments. Many of these other contributions are based on solid, if partial, historic datasets (e.g. Gauthiez, Zeller, and Guàrdia Bassols, this volume); however, for the archaeologist, the problems of particularism and context noted by these other contributors are multiplied and exacerbated by both temporal and socio-cultural distance. For most archaeological studies there are no texts, images, or footprints to aid reconstructions of the peopled cities of the past. The unit of analysis is thus the plano-metric representation of architecture and space (the modern Western map): a representation that has been critically analyzed in this volume. And yet, it is only through archaeology and the use of such representations that we are able to stretch our analyses back, beyond the Age of Exploration, over the eleven-or-so millennia of city life, and thus bring understanding from a variety of cultural groups. In this paper, we wish to explore the ancient Maya cityscape (or mapscape, as the case may be) of Copan, Honduras, and discuss some of the methods, theories, and models that we are employing to map not only architecture and space in the material sense, but also the dynamic changes that occur over time and through the movement of people. This paper is intended more as an exploration of method and consequence than as an interpretive work in its own right. In this light, we hope that the reader will agree that bringing life back to these long-dead cities opens the door for some extremely interesting and necessary discussions surrounding the archaeological record and its resulting representation.

2 Four Complementary Methods for Modelling Pedestrian Movement

From an operational standpoint, we will discuss four related methods for modelling pedestrian movement at the city level: axial line analysis, depth analysis (in both cases, focussing specifically on measures of integration), agent simulation, and crowd dispersal modelling. The first three are drawn from the literature on space syntax: a graph-based method for analyzing the relationship between human societies and the spatial configuration of buildings and settlements. Pioneered by a small group of young architects working out of London nearly four decades ago (Hillier et al. 1976), the methods and theories of space syntax have proven to greatly influence urban planners and theorists, particularly in Europe (e.g. Bafna 2003; Chang 2002; Ferguson et al. 2012; Haq and Zimring 2003; Hargreaves 2004; Kelbaugh 2001; Kusumo and Read 2003; Turner and Penn 2002; see also Knoespel (2003) for a recent historic example). Over the past decade and a half, a number of archaeologists have similarly explored many of the ideas that have continued to be developed by the practitioners of space syntax, with varying degrees of success (e.g. Chatford Clark 2007; Collins 2010; Dawson 2002; Fisher 2009; Morton 2012; Morton et al. 2012; Peuramaki-Brown et al. 2014; Seibert 2006; Shapiro 1999; Troncoso 2008; van Dyke 1999; Vaquer and Nielsen 2011; see also Bowser and Patton 2004 and Brusasco 2004 for ethnoarchaeological applications). While a survey of this research reveals that many (though by no means all) of these efforts continue to rely heavily on the theoretical and methodological strategies outlined in several seminal publications, particularly Hillier and Hanson's (1984) *The Social Logic of Space*, much has changed in the intervening years (particularly in terms of archaeological articulations with theory, see Smith 2010), and the methods and theories of space syntax have considerably more to offer from a data/theory standpoint than are typically utilized. The fourth method, crowd dispersal modelling, complements the aforementioned methods and, in many ways, is the easiest to apply (at least with regard to the simplified numerical calculations applied in this paper), though it lacks much of the theoretical foundation that has been constructed in support of space syntax. None of the approaches explored in this paper require any particular degree of technical aptitude; however, all require a critical approach to their application.

2.1 Axial Line Analysis

We have discussed the application of axial line analysis and depth analysis to archaeological contexts (their strengths and weaknesses) at length in previous publications (Morton 2012; Morton et al. 2012; Peuramaki-Brown et al. 2014). In our recent paper published in the *Cambridge Archaeological Journal*, we applied the methods of both axial line analysis and depth analysis to the city plan of Teotihuacan, Mexico, in order to explore structural similarities and differences at the individual apartment compound and broader city levels. While we will not explicitly revisit this analysis or its specific methods in detail, it should serve to briefly introduce both axial line analysis and depth analysis as applied to a best-case scenario.

The principal unit of space in an axial line analysis is the "grid": the pattern of linked spaces that define the system to be analyzed (Hillier 2002: 153). The particular aspect of the grid that we are interested in is the "structure": the pattern revealed by expressing the grid as an axial map and analyzing it configurationally (Hillier 2002: 153). The building blocks of this analysis are "convex spaces" and "axial lines": spaces in which all locations within them are mutually visible, and lines of visibility (or straight lines of potential movement) linking more than one convex space, respectively. There are a number of related measures within space syntax, discussed in the following paragraphs, which may be applied to this structure.

"Connectivity" (Hillier 1996: 94) indicates, quite simply, the number of immediately adjacent convex spaces to which any one space may be directly connected. Imagine that the room you are currently occupying is directly connected to three other rooms by three independent doors; thus, the room that you are currently occupying would have a connectivity of "3".[1] In the case of an axial line analysis, connectivity refers not to the number of immediately adjacent spaces, but to the number of other axial lines crossed by any particular line in question. As a general principle, relatively high connectivity is an important ingredient in determining how "busy" or "quiet" any particular space is likely to be (something we will return to shortly).

[1] If your imaginary room were directly connected to three rooms by four or five doors (or 52 doors for that matter), the room would still have a connectivity of 3.

"Line Length" (Hillier 1996: 142) is another relatively simple measure. Referring to axial lines, this measure quantifies the relative lengths of individual axial lines within a system. In general, longer lines of travel are likely to be better connected within the grid than shorter lines.

"Integration", or "Total Depth"[2] (Hillier 1996: 25), is one of the most fundamental measures in space syntax and indicates the minimum number of adjacent convex spaces, or axial lines, that one must traverse in order to move from any one space to all the others in the system. At a fundamental level, this measure is tied closely to both line length and connectivity as long, well connected, and typically, central axial lines are likely to be highly integrated. For the purposes of our stated task in this paper, "examining the human pulse that animated the ancient city", integration is particularly important. Within the space syntax literature, the structure of the grid has independent and systematic effects on movement patterns. This "Law of Natural Movement" (Hillier 2002; Hillier et al. 1993) states that the degree of integration of a space can be used as a predictor for how "busy" or how "quiet" that space will be (i.e. spaces with a lower total depth are "more integrated" and are likely to be "more busy", see Dawson 2002: 471; Peponis and Wineman 2002: 271).

Turning back to our best-case scenario, using the map of epicentral Teotihuacan produced by the Teotihuacan Mapping Project (Millon et al. 1973: map 2), a base map was produced that identifies the grid to be analysed (Fig. 1a). White space on this map indicates areas of the spatial grid that are interpreted as having been open to public traffic. Temple platforms, apartment compounds, boundary walls, and bodies of water are marked as inaccessible by black blocks or lines. An all-line axial analysis was then completed for this grid.[3] Our analyses suggested, perhaps not surprisingly, that the major axis of the Street of the Dead and the large eastern and western streets that bisect it, are the most strongly integrated paths in the urban grid (represented in Fig. 1b by a greyscale gradient in which the most highly integrated lines are represented as white, blending to black as integration decreases).

Teotihuacan is particularly amenable to such architectural analyses: not only is architectural preservation such that the urban grid can be reconstructed with sufficient accuracy to support the analysis, but the city grid itself is adequately "structured" or "constrained" to ameliorate the "boundary effects" that are apt to clutter the analysis of more "open" systems (e.g. most Maya centres, an effect that will be explored below). Upon cursory examination, Teotihuacan appears to share many more similarities (at least, in a superficial spatial sense) with modern urban centres (the systems with reference to which space syntax was initially formulated) than do those of the Classic Maya. We must therefore ask whether it is in fact

[2] In Hillier (1996), this measure is referred to as "mean depth"; however, the term is somewhat confused in the literature and as such, we have chosen instead the term "total depth" while the analysis remains the same.

[3] The analysis was conducted using a program called Depthmap, available for free academic download on the Space Syntax Laboratory website (www.spacesyntax.org).

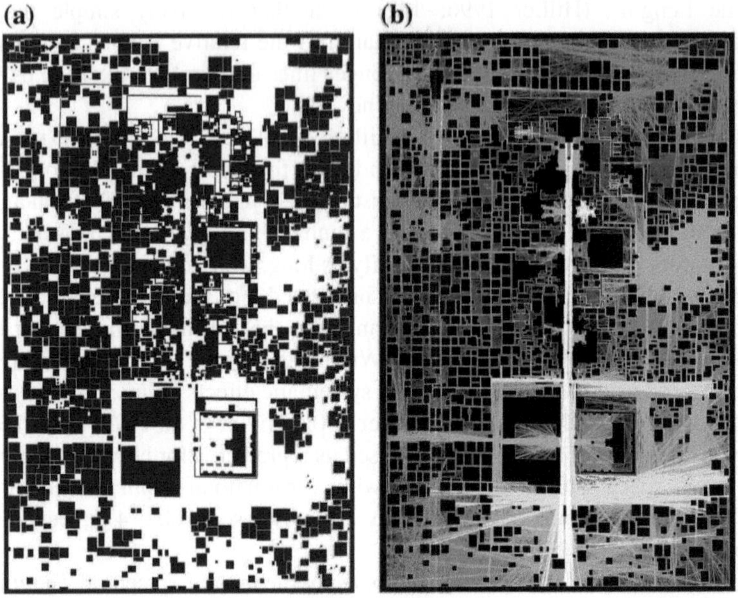

Fig. 1 **a** Map of epicentral Teotihuacan. **b** Axial analysis of Teotihuacan

possible to apply the methods of axial line analysis and the associated body of theory to explore the urban grid of Maya centres as well.

We believe that we can apply such methods, although some sites are more amenable to these analyses than others. At Teotihuacan, the formal structure of the grid itself served to contain and bind the analysis. In a publication produced for the *Canadian Journal of Archaeology* (Morton 2012; see also Morton 2007; Peuramaki-Brown et al. 2014), we applied an axial line analysis to the site of Naachtun, Guatemala. As with the majority of Maya cities, the architectural layout of Naachtun is open: few if any of the recognizable structuring features (i.e. roads and regularly-spaced structures) characteristic of modern urban forms (or indeed, characteristic of Teotihuacan) are present. Most sites consist of networks of organized plazas and disorganized, open, peripheral spaces. An axial analysis of such a system (Fig. 2a) illustrates the aforementioned "boundary effect": the longest, best-connected axial lines within the system are located around the margins of the monumental core (i.e. outside our intended subject of analysis). In order to profitably apply an axial line analysis to a site like Naachtun, we must somehow "bind" the analysis (Fig. 2b): an inherently interpretive act. Some boundaries may be social; for example, at Naachtun (as at Teotihuacán) we restricted the analysis to a public or commoner level by excluding platforms and structure interiors from the analysis. This is in contrast to the work of Seibert (2006) who explicitly sought to examine the relationship between these public and private spaces through the inclusion of structure and group interiors. Other

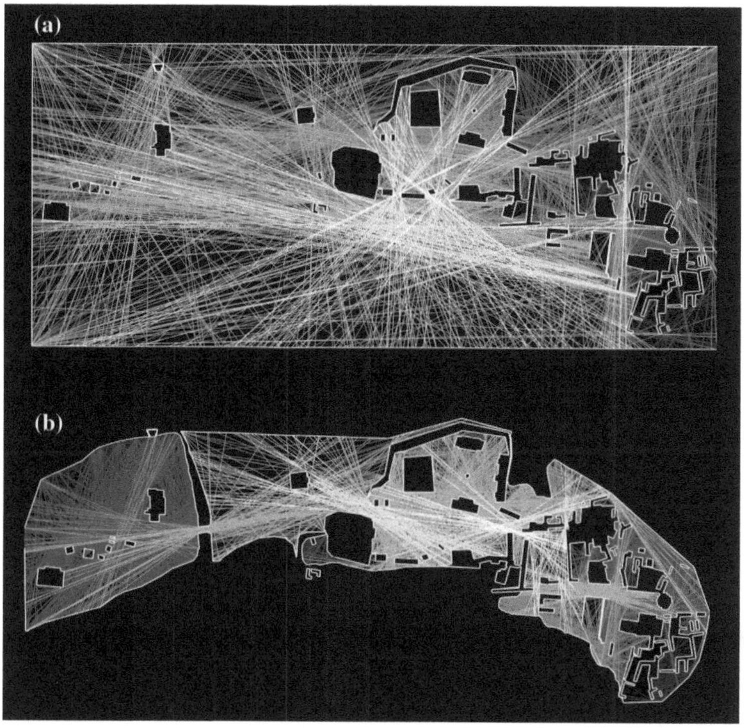

Fig. 2 **a** Unbound axial analysis of Naachtun. **b** Bound axial analysis of Naachtun

boundaries may be more practical; for instance, where steep or wet (but otherwise passable) terrains likewise serve as bounding features. In this manner, we force the analysis to conform more closely to the ways we expect people to have interacted with such spaces, while artificially forcing the analysis to focus on the lines of movement of most interest to us.

Thankfully, not all sites require such extreme manipulation in order to support axial analyses. At sites such as Becan (a moated site in Campeche, Mexico), Cahal Pech (a spatially-restricted hilltop site in Belize), or Copan, Honduras, sufficient structure is offered by the site grid to support the analyses with little requirement of artificial bordering. Applying an axial line analysis to the monumental centre of Copan (Fig. 3), we find little evidence for the boundary effects seen at Naachtun; rather, the analysis is simply bound by limiting it to the large open plaza (the Great Plaza) in the north of the civic-ceremonial centre and the increasingly restricted courtyards (the West and East Courts) to the south. Perhaps not surprisingly, the paths of highest integration are found in the Great Plaza. Below, we explore the implications of this observation.

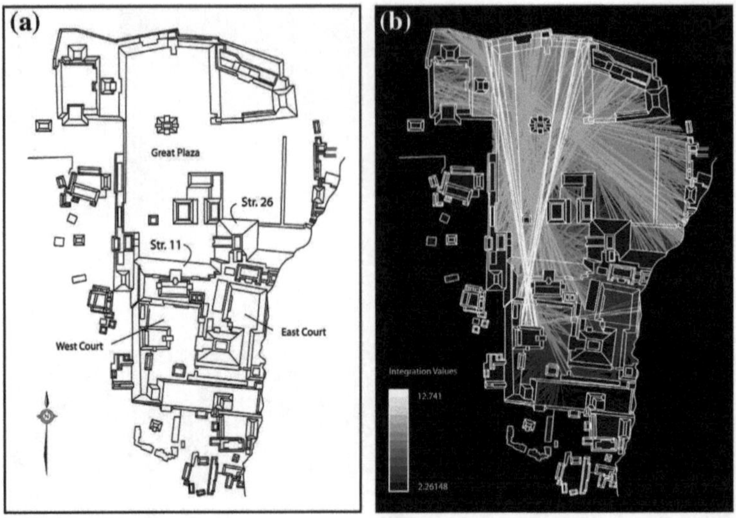

Fig. 3 **a** Map of Copan. **b** Axial analysis of Copan

2.2 Depth Analysis

A parallel depth analysis of convex spaces at Copan was conducted and provides additional insight to our discussion. We have already noted a number of characteristics of "Total Depth" where they overlap with axial line analysis. In traditional space syntax literature, there is often a methodological division between the two forms of analysis based on the types of systems being analyzed. Typically, axial line analysis is applied to external spaces and larger systems (city streets, shopping centres, etc.). Depth analyses, as in this study, are more commonly applied to internal spaces and smaller systems (houses, office buildings, etc.). We previously applied such an analysis to a number of apartment compounds at Teotihuacan (Morton et al. 2012). Using the Zacuala Palace as an example, we divided the rooms of the compound into a series of distinct convex spaces and calculated the total depth of each space of interest. In Fig. 4, the depth counts for one such space (shaded grey) is shown with the "total depth" being the sum of the minimum number of steps required to pass from the carrier space to each other space in the system.

As noted above, Maya urban centres, including those that are spatially restricted such as Copan, do not conform to the typical city grid to which axial line analysis is typically applied. In many ways, such centres appear much more spatially analogous to structures/spaces such as Teotihuacan's apartment compounds or to modern architectural interiors. Applying such an analysis to the monumental core of Copan (Fig. 5) produces similar results to those produced by the axial analysis (with the lowest total depths being found in the Great Plaza), albeit with several interesting differences that will be explored below.

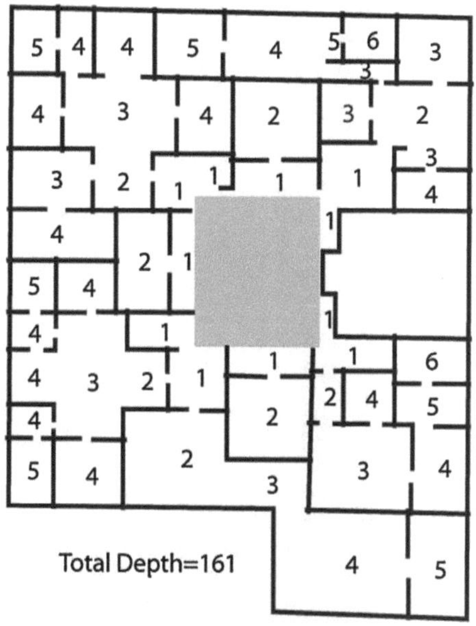

Fig. 4 Example of depth analysis at Zacuala palace, Teotihuacan

2.3 Agent Simulation

A related technique is Agent Simulation Modelling (ASM). ASM programs are designed to predict human movement within built environments and landscapes. Depending on the sophistication of the particular model employed, it is possible to simulate a variety of behaviours that vary according to the specific situation modelled (including the goals of individual or groups of agents), the size of the encounter modelled, and the user application adopted. ASM is widely applied in modern urban planning, not only for analytical purposes but also as a powerful visualization tool. Some simulations are incredibly complex and well beyond the scope of this paper. For now, we do not need to stray too far afield from our work with space syntax. Conveniently, Depthmap also contains a series of simple tools that can be of use for basic agent simulation. The agent model for Depthmap has somewhat different theoretical foundations than the space syntax model (though roughly analogous); nonetheless, it shares a fundamental emphasis with space syntax and axial line analysis in that the primacy of the visual field is what influences human movement.

In terms of the analysis, ordinarily the site plan in question is divided up into a more-or-less arbitrary grid, and a full 360-degree visual sweep is split up into 32-evenly-divided angular sections, or "bins" (each 11.25° wide). The default settings in the analysis restrict this to 15 bins (a total 168.75°) arrayed in front of the agent, approximating a standard human visual field (with slight back-and-forth movements of the head). From these, the agent decides which paths to follow. Gate counts on each grid space record the number of times that it is crossed by an agent.

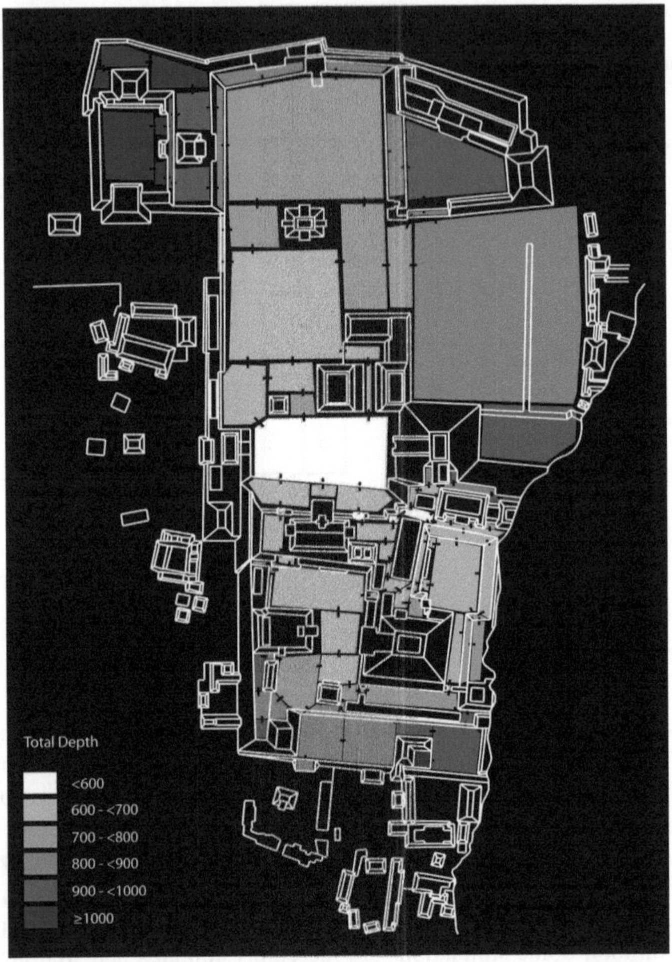

Fig. 5 Depth analysis at Copan

Following the broader rules of space syntax, agents populating our spatial grid are attracted to more expansive/longer visual fields and favour linear paths of movement to those that double back (i.e. changes in path are more likely when the change can be made at an obtuse angle rather than an acute). Other view fields produce different results; for instance, lowering the bin number (e.g. down to five) seems to encourage agents to follow exceedingly straight trajectories (and produces generally equal gate counts throughout the system), while raising the bin number (e.g. up to 32) seems to encourage the agents to cluster in open areas near where they spawned (with correspondingly higher gate counts in these areas). Raising and lowering the "steps before turn decision" seems to affect the simulation in the same way, with three steps being taken as optimal. Research by Al

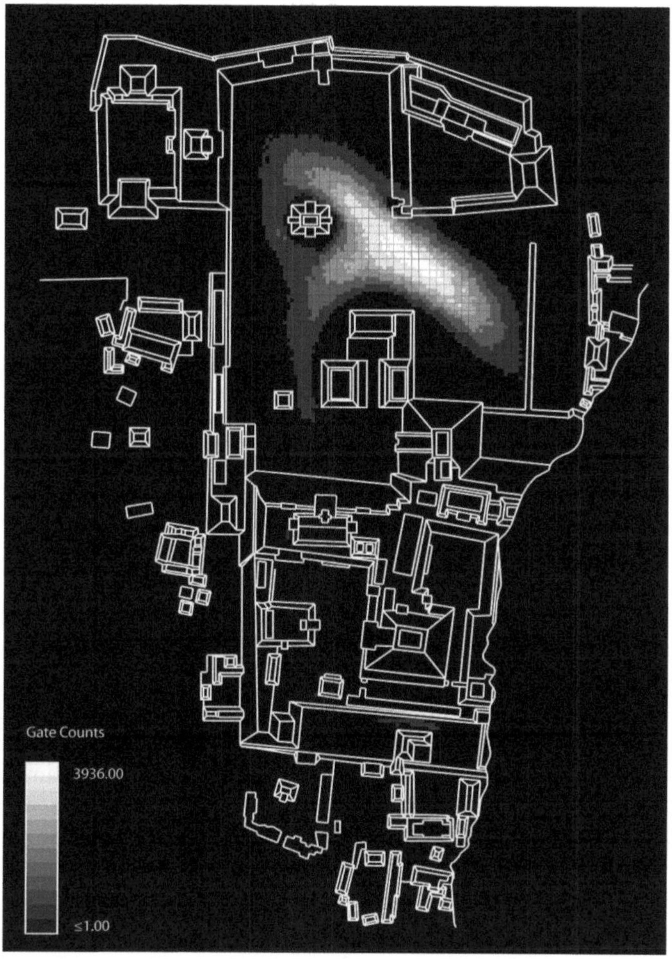

Fig. 6 Agent simulation at Copan. *Note* great Plaza is unobstructed

Sayed and Turner (2012) and Turner (2004) indicates that Depthmap's default settings seem to better model cross-culturally observed, non-purposive, movement than do other possible combinations.

An analysis of Copan's site core (Fig. 6), in which 10,000 agents were randomly released within the grid and their movements tracked over a thousand "steps" each,[4] shows a similar pattern to that seen in the axial line and depth analyses, with one major exception: while the axial line analysis may be useful for predicting potential lines of travel that may have attracted movement, and the

[4] Agent movement, in this case, is non-purposive (i.e. not goal oriented). Agents simply explore the urban grid of Copan for the duration of the analysis rather than trying to move from any particular space to any other.

depth analysis models how "busy" or how "quiet" any particular convex space was likely to be, the agent simulation models how people may have actually moved through these spaces. There is analytical power in this and we will explore a number of insights afforded by the analysis (or more properly, *for* the analysis) in the following section.

2.4 Crowd Dispersal Modelling

We would end this exploration of method with a final, complementary technique for modelling human/space interaction. To this end, we move away from space syntax, axial line and depth analysis, and Depthmap.

There are generally two applications for crowd simulations. The first involves the use of photorealistic renderings of crowd scenes in Hollywood films, video games, and virtual heritage environments (Thalmann et al. 2009). The second is focused on developing effective crowd management scenarios during emergencies (Shendarkar and Vasudevan 2006). Both seek to simulate the behaviour of real crowds as accurately as possible, often using cellular automata or agents. At a very basic level, crowds behave in ways that are similar to fluids and gasses (Luo et al. 2009: 275); consequently, early studies utilized particle simulation models to predict the flow of human movement in various environmental settings. More recently, psychological and sociological models have been incorporated into crowd simulations (Helbing and Molnar 1998). Factors such as aggressive and non-aggressive behaviour, stimulated by collision-avoidance responses among agents, have proven effective in understanding human responses to changes in crowd density as well as environmental boundaries such as walls and narrow exits (Wang et al. 2012; Turkay et al. 2011). Advances in artificial intelligence (AI) allow for even more complex social forces to be explored. By way of illustration, managing human flow is often complicated by the fact that individuals are simultaneously trying to achieve their own objectives, frequently at the expense of other agents.

The use of Belief-Desire-Intent (BDI) architecture provides a means of simulating these forms of complex scenarios. BDI architecture imbues simulation agents with a mental architecture that allows them to behave in ways that are autonomous, cooperative, learnable, and adaptable (Shendarkar and Vasudevan 2006: 546). Agents can also be assigned any number of unique characteristics such as age, sex, knowledge of area, panic scale, leadership, interdependence, and injury scales (Helbing and Molnar 1998). These types of advanced scenarios have been used to examine the underlying causes behind some of the worst crowd disasters in recent history, including the Glasgow iBrox Stadium accident in which 66 people were killed; the 1996 Guatemala City tragedy in which 84 people died on a football field; and the Beijing Miyun disaster in which 37 people died in a stampede along the Mihong Bridge (Wang et al. 2012: 1). Simulations have demonstrated that crowd characteristics such as density, diversity, speed of movement, and motivation of individuals played a large role in these disasters.

While we have yet to attempt to produce these complex forms of simulation, existing research provides us with many insights into simulation variables that archaeologists need to consider when attempting these forms of study. Crowd diversity, density, speed, and direction of movement all play critical roles in simulation research. Crowd diversity includes such factors as age, sex, levels of aggression, and physical size (Wang et al. 2012; Turkay et al. 2011; Shendarkar and Vasudevan 2006). With respect to the latter, current crowd evacuation research has shown that body thickness and shoulder breadth of normal adults can play a role in determining the number of individuals that can comfortably occupy a unit of space (Wang et al. 2012: 6). This measure is called maximum bearing density, and it demonstrates that crowd density can vary independently of its physical size. For example, the maximum bearing density in England can been calculated to be 7.7 people/m^2 while in Japan it is 10.3 people/m^2 (Wang et al. 2012: 7). Consequently, estimates of the maximum physical size (body thickness and shoulder breadth) of past populations like the Maya may be necessary to produce accurate crowd simulation models.

Crowd density is perhaps the key to understanding crowd dynamics, as it directly influences speed and direction of movement (Wang et al. 2012; Turkay et al. 2011; Shendarkar and Vasudevan 2006). Generic simulations demonstrate that when crowd densities are low, pedestrians can move in a variety of directions at free walking speeds of 1.3 m/s (Wang et al. 2012: 4). From the perspective of social forces, collision-avoidance with other pedestrians and environmental borders is facilitated under these conditions, thereby reducing aggression among agents (Helbing and Molnar 1998). As crowd densities begin to increase, crowd speed is diminished whereas crowd flow continues to increase; for example, free walking speeds are attained at crowd densities of between 0.2 and 0.27 people/m^2. Crowd flow will continue to increase until a critical value of around 4.5–7 people/m^2 is reached, at which point movement stagnation occurs (Wang et al. 2012: 4). At this stage, the occurrence of a critical incident such as a person falling down or an escape route becoming blocked, can cause the crowd to easily collapse, resulting in disaster. Interestingly, crowd simulation research shows that the number of exits available during an emergency evacuation is often inconsequential because no crowd flow can be formed when crowd densities are more than 5 people/m^2 (Wang et al. 2012: 8); thus, it may not be necessary for archaeologists to determine all points of entry/exit from an area under study. Of the techniques explored in this paper, it is perhaps this last that may prove the most intriguing. Next, we will explore the particulars of this type of analysis as it applies to Copan.

3 Putting it All Together: Interpretation at Copan

The goal of this paper is to explore the complex entanglements that define urban spaces through the various lenses afforded by a reconsideration of the limitations imposed by, and opportunities fostered through, the analysis of mapped space (i.e. how our analyses are tied to the way we represent space). Other papers in this volume have discussed similar issues from a more theoretical perspective and have questioned the validity of the Western tradition of spatial representation. While we acknowledge these same limitations, it is our goal to demonstrate the nuanced methods that may nonetheless be applied to, and interpretations that may nonetheless be drawn from, such traditional representations (albeit, with application of a critical hand). Above, we discussed in brief the application of a number of techniques for analyzing the spatial form of urban centres, and applied these analyses to the ancient Maya site of Copan, Honduras. We will conclude this paper with an exploration of the socio-spatial interpretations and, perhaps more significantly, the new questions that may be generated through such applications (both of the archaeological record and our methods of representation).

Each of the three related methods for analyzing urban form from a space syntax perspective offers different but complementary insights. Where these analyses agree is in the identification of the Great Plaza as the busiest, or most heavily trafficked, portion of the site as a simple function of the urban grid itself. This is hardly an earth shattering observation, and has long been interpreted as such (e.g. Gordon 1899). Rather, where our analyses differ from one another is where the greatest insight is gained: each highlights slightly different areas of the Great Plaza based on the particular qualities of modelled pedestrian movement. These differences offer interesting interpretations/questions in their own right, while also pointing to a number of significant consequences of the traditional form of planometric mapping as it is particularly applied in Maya archaeology.

The depth analysis (Fig. 5) serves as a particularly potent framework within which to interpret the other two analyses. The depth analysis shows unambiguously that the areas of lowest total depth are those immediately in front of Str. 11 and the Str. 26 Hieroglyphic stairway, not to mention on the stairs of Str. 11 itself. The analysis does not seek to interpret specific paths of movement *per se*, but rather predicts how "busy" or "quiet" any individual convex space is likely to be. What the depth analysis highlights is the degree of spatial control exerted by the restricted access points to Copan's upper courts, effectively bisecting the site in two and presumably in reflection of very real social divisions: the West and East Courts (the restricted courtyards on the southern end of the site) are surrounded by smaller temple structures and élite residences/administrative buildings. Not only is space in these areas restricted, thereby limiting the number of individuals that could occupy them at any one time, they are also both elevated over the larger Great Plaza to the north. Together, these characteristics may have fostered a sense of exclusivity (a pattern also noted at other sites, see Andres et al. 2010; Awe 2008; Awe et al. 1991; Hammond 1972) while also serving as an unambiguous symbolic tie between the open common space (north) and the exclusive élite space (south): hierarchical divisions writ large in the generations-old architecture of the urban core.

The axial analysis (Fig. 3), on the other hand, highlights the integrative qualities of specific paths of movement between convex spaces, in particular identifying the western half of the Great Plaza through to the West Court as the most heavily integrated paths (those most likely to encourage pedestrian movement). However, during Copan's apogee in the decades following the early 8th century rule of 18 Rabbit (*Uaxaclajuun Ub'aah K'awiil*), it is likely that access to these upper areas of the site core and their associated architecture were heavily restricted: if not from a spatial standpoint, then from a social one. Experientially, this particular social boundary (the shallow space in front of Str. 11) may have been rather acutely felt as the connection between the Great Plaza and the upper courts would have been a natural one, though artificially stifled; however, there remains cause to be cautious in making such interpretations.

As previously noted, the agent simulation (Fig. 6) operates on a slightly different set of formal criteria from the other syntax analyses. While the results generally accord well with the axial and depth analyses, where they differ are in the details. In this analysis, given a random origin for the agents, there is very strong clustering (observable as high gate counts) in the eastern portions of the Great Plaza. This is at least in part an artifact of the analysis itself: given a random point of origin for the agents in the simulation, there is likely to be more appearing in larger open spaces than in smaller restricted ones. What is more significant is that individual agents do not seem to follow the paths of highest integration as identified by the axial analysis.

This discrepancy highlights a serious flaw in the way we typically apply space syntax at the city level, and results from the way we traditionally perceive mapped space. The "God's View" of large urban spaces that is typical of Western representation can have a tendency to remove the human from the city. This does not mean that broader analyses are irrelevant. The spatial characteristics identified in the axial analysis (in this case) lie at the heart of the agent simulation that helps us understand the complicated ways that people may have actually interacted with the spatial grid in question: while agent movement is dictated by view field (as in the axial analysis), agents are programmed to follow a more "natural" meandering path, turning slightly every few steps and reassessing their view field at each point. Thus, as an agent progresses along the long vistas that are thought to attract movement in the axial analysis, they are simultaneously and with each step, reducing the scope of the view field that had attracted them in the first place. Eventually, other paths seem to preferentially draw their attention. Agents seem to cluster somewhat within larger open spaces and only rarely venture through the more restricted upper courts. With this information in hand, we might mollify our initial interpretation from the axial analysis that suggested a fundamental and acute tension between social classes fostered by the disjunction between the natural paths of movement identified in the analysis, and by the assumed social boundaries that would have restricted access to the upper courts. Our crowd dispersal analyses also benefit from a comparative approach.

The crowd dispersal analyses outlined above are the most provocative from a human/space perspective, and highlight a difficult series of questions for archaeological interpretation. Using simple measurements from the Copan site map, we can apply some of the basic observations from the generic crowd simulations discussed in the previous section. It is estimated that 15,008 people occupied

Copan in AD 749 (a decade after 18 Rabbit's death), with population levels rising to 27,753 by AD 799 (Paine et al. 1996; Webster et al. 1992). Further, it has been suggested that the plenum of the population occasionally gathered for important socio-political rituals within Copan's site core, specifically centred in the Great Plaza. Although such events have been interpreted as extremely important to the proper functioning and legitimisation of the State, archaeologists have not examined the potential consequences of such gatherings from a crowd control perspective.

Knowing the size of the Great Plaza at Copan (35,800 m^2), a dimension that remained static despite the need to accommodate a bloating population, allows us to calculate the crowd density for each of these periods. At AD 749 we can calculate a maximum crowd density of 0.4 people/m^2, rising to 0.8 people/m^2 in the later period of highest occupation. Given that free walking among pedestrians occurs at crowd densities of 0.2–0.27 people/m^2, we can speculate that ceremonial events occurring in the Great Plaza would have mildly restricted multi-directional flow and required some measure of collision-avoidance among attendees. The degree to which this would have affected the character of the crowd is unknown though it would seem that pedestrian flows into and out of the Plaza could have been maintained under these crowd densities.

It is at this point that the perceived spaces as traditionally represented, may be substantially misleading. Recent work at other ancient Maya centres has suggested that plaza spaces may not have been the open, empty, flat areas that have typically been assumed. At the site of Buenavista del Cayo, Belize, Capp (2013) has demonstrated that the East Plaza was cluttered with both semi-permanent activity areas and more substantial masonry foundations for perishable superstructures. At the sites of Tipan Chen Uitz, Yaxbe, and Cahal Uitz Na, all in central Belize, plazas appear to have been only roughly finished, with bedrock outcrops and associated terracing being common. From an analytical standpoint, the consequences of these obstacles may be extremely relevant; for example, what is the effect of imposing a randomly distributed series of obstacles in the Great Plaza? In the case of the agent simulation pictured in Fig. 7, though only 20 % of the plaza space was directly impacted, the potential effects on movement within the system are staggering. The simulation demonstrates a much more evenly distributed series of gate counts than did the open simulation pictured in Fig. 6. While the interpretive consequences of this remain to be explored in another venue, here we would focus on the methodological consequences.

These factors may be directly significant in consideration of our crowd dispersal data. Simulations show that crowd densities of between 7 and 13 people/m^2 constitute extreme situations that can lead to movement stagnation, injury, and death. Although in this case only 20 % (or ~7,160 m^2) of the Great Plaza was directly cancelled by the hypothetical presence of structures (in absolute terms, only raising the maximum crowd densities to 0.5 people/m^2 and 1.0 people/m^2 for early and late periods, respectively), such structures may have had even greater effects by simultaneously cancelling space and parcelling/blocking other areas; for instance, if the structures and impediments as modelled in Fig. 7 effectively

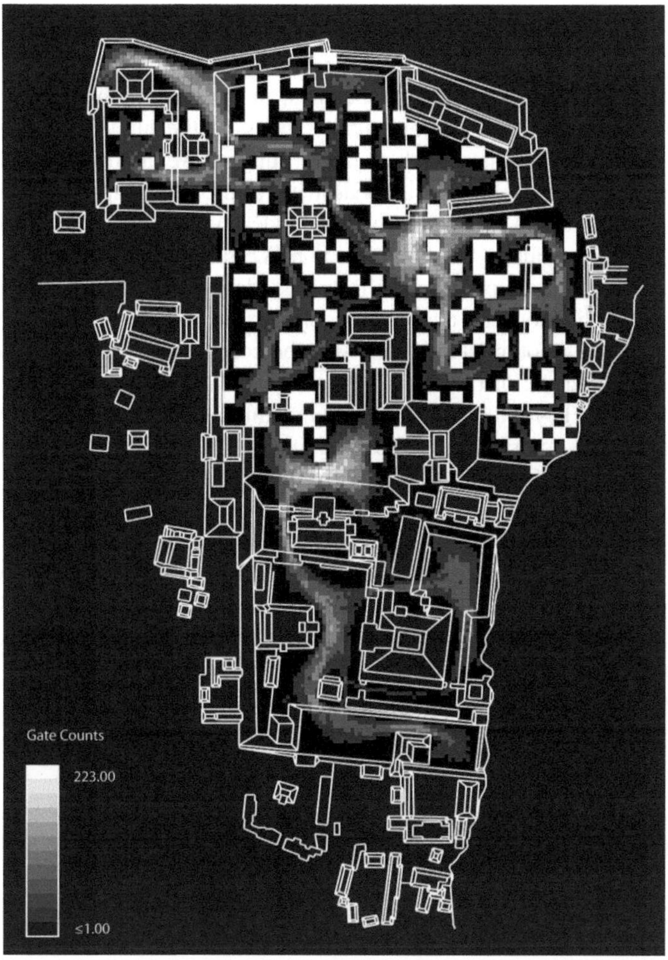

Fig. 7 Agent simulation at Copan. *Note* great Plaza with 20 % random distribution of obstacles

segregated large portions of the plaza, then the effects on crowd density (and the axial analyses completed earlier, for that matter) may have been significant. For now, it is sufficient to note that, depending on the degree to which permanent or semi-permanent obstacles were present, dangerous overcrowding was perhaps an issue faced by later planners in Copan. If so, then events similar to the Mihong Bridge stampede and the 2010 Khmer Water festival disaster may also have occurred in antiquity. Being the largest architectural element at most sites, it would be logical to devote more time to the exploration of the plaza.

4 Final Thoughts

We hold that the study of complex societies, in particular urban spaces such as those of the ancient Maya, can effectively focus on the human interactions and entanglements that animated such locales. Further, we hold that many of the concerns related to crowd dispersal, pedestrian traffic patterns, the constitution of community, and socio-spatial control that underlay spatial analyses of modern urban centres were equally valid in past, non-Western, urban centres. The advent of computer simulation programs allows the development and exploration of models and implications of ever more complex hypotheses concerning human behaviour, enabling potential simulations to be compared with the archaeological record. Archaeologists, in particular Mayanists, have been slow in adopting any simulation models beyond 3D virtual reality settings; thus, presenting architectural stages without the actors. This is unfortunate as the continual development and advances in technological simulation are allowing the input and integration of an increasing number of data sets: what should be viewed as an extremely useful tool in a field where practitioners are constantly struggling to organize and compare information. However, these techniques should not be applied uncritically. Hence, in this paper, and in concert with the other papers presented in this volume, we have paid particular attention to the analytical consequences (both opportunities and limitations) of such mapping. These forms of analysis are still very much in their infancy in Maya archaeology; however, we believe they represent a strong and viable avenue for future research.

References

Al Sayed K, Turner A (2012) Agent analysis in depthmap 10.14: manual. University College London, London

Andres CR, Wrobel GD, Morton SG (2010) Tipan Chen Uitz ("fortress mountain well"): a major "new" Maya center in the Cayo district, Belize. Mexicon 32:88–94

Awe JJ, Campbell MD, Conlon J (1991) Preliminary analysis of the spatial configuration of the site core at Cahal Pech, Belize and its implications for lowland Maya social organization. Mexicon 8(2):25–30

Awe JJ (2008) Architectural manifestations of power and prestige: examples from classic period monumental architecture at Cahal Pech, Xunantunich, and Caracol, Belize. Res Rep Belizean Archaeol 5:159–173

Bafna S (2003) Space syntax: a brief introduction to its logic and analytical techniques. Environ Behav 35(1):17–29

Berger J (1992) Keeping a rendezvous. Vintage International, New York

Bettencourt L, West G (2010) A unified theory of urban living. Nature 467:912–913

Bettencourt LMA, Lobo J, Helbing D, Kühnert C, West GB (2007) Growth, innovation, scaling, and the pace of life in cities. PNAS 104(17):7301–7306

Bettencourt LMA, Lobo J, West GB (2009) The self similarity of human social organization and dynamics in cities. Complexity Perspect Innov Soc Change 7:221–236

Bowser BJ, Patton JQ (2004) Domestic spaces as public places: an ethnoarchaeological case study of houses, gender, and politics in the Ecuadorian Amazon. J Archaeol Method Theory 11(2):157–181
Brusasco P (2004) Theory and practice in the study of Mesopotamian domestic space. Antiquity 78:142–157
Capp B (2013) Classic Maya economies: a study of marketplace exchange at Buenavista del Cayo, Belize. Unpublished Ph.D. Dissertation, Department of Anthropology, University of Wisconsin, Madison
Chang D (2002) Spatial choice and preference in multilevel movement networks. Environ Behav 34(5):582–615
Chatford Clark DL (2007) Viewing the liturgy: a space syntax study of changing visibility and accessibility in the development of the byzantine church in Jordan. World Archaeol 39(1):84–104
Collins WM (2010) The center as cultural and grammatical theme in Mam (Maya). Space Cult 13(1):17–31
Dawson PC (2002) Space syntax analysis of central inuit snow houses. J Anthropol Archaeol 21(4):464–480
Ferguson P, Friedrich E, Karimi K (2012) Origin–destination weighting in agent modelling for pedestrian movement forecasting. In: Proceedings of the international space syntax symposium, vol 8: 8153:1–8153:13
Fisher KD (2009) Placing social interaction: an integrative approach to analyzing past built environments. J Anthropol Archaeol 28:439–457
Gordon GB (1899) The ruined city of Copan. J American Geogr Soc New York 31(1):39–50
Hammond N (1972) The planning of a Maya ceremonial center. Sci Am 226(5):82–91
Haq S, Zimring C (2003) Just down the road a piece: the development of topological knowledge of building layouts. Environ Behav 35(1):132–160
Hargreaves A (2004) Building communities of place: habitual movement around significant places. J Hous Built Environ 19(1):49–65
Helbing D, Molnar P (1998) Social force model for pedestrian dynamics. Institute of Theoretical Physics, University of Stuttgart, Germany
Hillier B (1996) Space is the machine. Cambridge University Press, Cambridge
Hillier B (2002) A theory of the city as object: or, how spatial laws mediate the social construction of urban space. Urban Des: Int 7:153–179
Hillier B, Hanson J (1984) The social logic of space. Cambridge University Press, Cambridge
Hillier B, Leaman A, Stansall P, Bedford M (1976) Space syntax. Environ Plann B-Plann Des 3:147–185
Hillier B, Penn A, Hanson J, Grajewski T, Xu J (1993) Natural movement: or, configuration and attraction in urban pedestrian movement. Environ Plann B: Plann Des 20(1):29–66
Kelbaugh D (2001) Three urbanisms and the public realm. In: Proceedings of the international space syntax symposium, vol 3, pp 14.1–14.8
Knoespel KJ (2003) Building space and myth at the edge of empire: space syntax analysis of St. Petersburg, 1703–1913. In: Proceedings of the international space syntax symposium, vol 4, pp 17.1–17.14
Kühnert C, Helbing D, West GB (2006) Scaling laws in urban supply networks. Physica A 363(1):96–103
Kusumo C, Read S (2003) Building on geometries of intelligibility: planning the Leiden region as a layered movement fabric. In: Proceedings of the international space syntax symposium, vol 4, pp 18.1–18.14
Luo L, Zhou S, Cai W, Low MYH, Michael L (2009) Toward a generic framework for modeling human behaviors in crowd simulation. Paper presented at the 2009 IEEE/WIC/ACM international conference on web intelligence and intelligent agent technology. Milano, Italy
Millon R, Drewitt RB, Cowgill GL (1973) The Teotihuacan map. Part two: maps. Urbanization at Teotihuacan, vol 1. University of Texas Press, Austin and London, Mexico

Morton SG (2007) Procession ritual at Naachtun, Guatemala during the late classic period. Unpublished MA Thesis, Department of Archaeology, University of Calgary, Calgary

Morton SG (2012) Ritual procession and the creation of *Civitas* among the ancient Maya: a case study from Naachtun, Guatemala. Can J Archaeol 36(1):141–165

Morton SG, Peuramaki-Brown MM, Dawson PC, Seibert JD (2012) Civic and household community relationships at Teotihuacán, Mexico: a space syntax approach. Cambridge Archaeol J 22(3):387–400

Paine RR, Freter AC, Webster DL (1996) A mathematical projection of population growth in the Copan valley, Honduras, A.D. 400–800. Latin. Am Antiq 7(1):51–60

Peponis J, Wineman J (2002) Spatial structure of environment and behavior. In: Bechtel RB, Churchman A (eds) Handbook of environmental psychology. Wiley, New York, pp 271–279

Peuramaki-Brown MM (2012) The integration and disintegration of ancient Maya urban centres: charting households and community at Buenavista del Cayo, Belize. Unpublished Ph.D. Dissertation, Department of Archaeology, University of Calgary, Calgary

Peuramaki-Brown MM, Morton SG, Dawson PC, Seibert JD (2014) The dynamic Maya city: methods for modelling pedestrian movement in ancient civic-ceremonial centres. In: Kettunen H, Helmke C (eds) Proceedings of the 17th European Maya conference, Helsinki 2012. European Association of Mayanists, WAYEB, pp TBA

Scott JC (1998) Seeing like a state: how certain schemes to improve the human condition have failed. Yale University Press, New Haven and London

Seibert JD (2006) Classic Maya range structures and socio-political organisation. Unpublished Ph.D. Dissertation, Department of Archaeology, University of Calgary, Calgary

Shapiro JS (1999) New light on old Adobe: a syntax analysis of the Casa Grande. Kiva 64(4):419–446

Shendarkar A, Vasudevan K (2006) Crowd simulation for emergency response using BDI agent based on virtual reality. In: Proceedings of the 2006 winter simulation conference. Monterey, California, pp 545–553

Smith ME (2010) Empirical urban theory for archaeologists. J Archaeol Method Theory 18(3):167–192

Thalmann D, Grillon H, Maim J, Yersin B (2009) Challenges in crowd simulation. Paper presented at the 2009 international conference on cyberworlds. University of Bradford, UK

Troncoso A (2008) Spatial syntax of rock art. Rock Art Res 25(1):3–11

Turkay C, Koc E, Balcisoy S (2011) Integrating information theory in agent-based crowd simulation behavior models. Comput J 54(11):1810–1820

Turner A (2004) Depthmap 4: a researcher's handbook. University College London, London

Turner A, Penn A (2002) Encoding natural movement as an agent-based system: an investigation into human pedestrian behaviour in the built environment. Environ Plann B: Plann Des 29:473–490

van Dyke R (1999) Spaces syntax analysis at the Chacoan outlier of Guadalupe. Am Antiq 64(3):461–473

Vaquer JM, Nielsen AE (2011) Cruz Vinto from the surface: reach and limitations of space syntax in a settlement from later regional development period (ca. 1200–1450 AD) in Northern Lipez, Bolivia. Revista española de antropología Americana 41(2):303–326

Wang J, Lo S, Wang Q, Sun J, Mu H (2012) Risk of large-scale evacuation based on the effectiveness of rescue strategies under different crowd densities. Risk Anal 1–11

Webster DL, Sanders WT, van Rossum P (1992) A simulation of Copan population history and its implications. Ancient Mesoamerica 3:185–198

Mapping Socio-Spatial Relations in the Urban Built Environment Through Time: Describing the Socio-Spatial Significance of Inhabiting Urban Form

Benjamin N. Vis

Abstract This chapter introduces Boundary Line Type (BLT) mapping, a vector GIS based cross-culturally and diachronically comparative method, used for mapping the socio-spatial significance of urban built environments. This new research method is related to other methods currently used to study contemporary as well as historical urban built environments such as urban morphology, space syntax, and GIS based approaches. BLT mapping uses GIS technology in order to apply an ontology of formal boundary conceptualisations expressing the constitutive differences among the materially constructed subdivisions which shape built environments and are inhabited by urban society. This ontology resulted from a firm socio-spatial theoretical grounding (Vis in Sp Flows: Int J Urb ExtraUrb Stud 2(4): 15–29, 2013a; Vis 2013b; Vis in J Borderland Stud, forthcoming) and is here operationalised on the basis of contemporary, historical, historically reconstructed, and archaeological ground-level city plans of the historic city of Winchester (UK) and Chunchucmil (Classic Maya, Mexico). The research processes of data preparation and the analytical mapping of BLTs by identifying them in empirical data contexts are presented. This alerts the prospective user to the challenges and practical measures involved in using spatial datasets of different origin. The interpretive opportunities of the resultant formal redescription of the urban landscape and the potential of the BLT data structure for both advanced spatial analysis and visualisation is explained. Facilitating this interpretive and analytical mapping practice is expected to stimulate future research to systematically explore society-space relations as manifest and developing in cities over time and in socio-culturally contrasting urban traditions. Devising and conducting this methodology advances the qualitative GIS research agenda for the spatial humanities and social sciences by marrying theoretically informed ideational concepts to quantifiable empirical units of information.

B. N. Vis (✉)
School of Geography, University of Leeds, Leeds, England, UK
e-mail: B.N.Vis10@leeds.ac.uk

S. Rau and E. Schönherr (eds.), *Mapping Spatial Relations, Their Perceptions and Dynamics*, Lecture Notes in Geoinformation and Cartography,
DOI: 10.1007/978-3-319-00993-3_4, © Springer International Publishing Switzerland 2014

1 Introduction

Recent years have seen a distinct rise in the use of Geographical Information Systems (GIS) for the historical study of cities. The urban branch of the broader emerging practice of historical GIS, is taking advantage of the flexibility in data collation and integration offered by a digital database based environment over draughting plans on paper (Lilley, *Mapping Medieval Chester*). Despite GIS's geostatistical and geoanalytical underpinnings, historical GIS is aimed foremost at locating historical sources, data and events on a map to visualise them in spatial distributions, which may help to explain historical processes and relations (see Gregory and Ell 2007).

Urban examples of such digital mapping practices show a similar preference for locating the past on historical plans, often in relation to the current city plan (e.g. Frank 2013; Locating London's Past; Amherst College 2009; Amherst College 2010; Jensen and Keyes 2003). Many projects are not primarily concerned with creating a reconstruction of the urban environment that is as accurate and historically sensitive as possible. *Mapping Medieval Chester* (Faulkner), *Mapping Medieval Townscapes* (Lilley et al. 2005), *Pompeii Bibliography and Mapping Project* (Poehler), *Alpage* (Noizet and Costa 2013; Noizet and Grosso 2011) and a GIS for medieval Antwerp (Bisschops 2012) are excellent examples of the meticulously detailed reconstructed plans that can be achieved using digital tools. These open up promising future directions for social and spatial research based on such data. Inspired by the possibilities enabled by historically regressive sequence mapping of cities, this paper aims not to present another example of this particular practice, but to introduce a new mapping method which exploits the best level of historically or archaeologically reconstructed spatial data that could be achieved for the comparative socio-spatial study of the physical characteristics of the built environment through time.

As demonstrated by historical GIS efforts, historians, akin to human geographers (see Jones et al. 2009), usually map to spatially visualise what happened where and investigate resultant locational relations. Instead, in social scientific archaeology or historical geography one might want to ask: *how* was it where something happened? This addresses the pertinence of revealing the affective qualities and affording particulars that characterise the location and spatial situation in which something occurred, next to questions that just determine where something occurred. Fletcher (2004) expressed a similar concern from the perspective of materiality in space, wanting to ask what it is the material does to create what actually happened in collision with social action. Both questions are part of studying the relationships between society and space, which according to Griffiths (2013) is one of the main reasons for historical science to engage with the geographical practice and theory of GIS. Furthermore, Griffiths (2013:154) argues, history's foci on the study of maps as cultural objects and historical space as representation have "created something of an epistemological blind spot for historians wishing to access and substantively describe 'spaces of practice' produced by everyday activity." The material implications of the proposed line of questioning, cf. the discipline of archaeology, would thus prepare the historical fields for uncovering the entanglement of living in the material spatial world over the long term. Such endeavour would contribute to answering Lilley's (2011b) alarm over historical geography's neglect of the Middle Ages and Jones' (2004) stressing of the benefits of extending human geography's temporal frame of reference beyond the recent past.

This perspective would enable one to explore and better understand how the experiential and affording[1] structure of an urban built environment developed as a social reality and what the compositional properties were in which this structure was originally encountered. Effectuating such research requires a general conceptualisation of the socio-spatial significance with regard to inhabiting and developing the built environment, incorporated in the basic physical characteristics composing urban landscapes. In other words, a conceptualisation making explicit the spatial and material features affording urban life and development to take place in a time–space specific way in each city. Formulating the theory and conceptualisation necessitated by this objective has been the subject of earlier research (Vis 2013a, b, forthcoming).

Founded on an interdisciplinary body of thought on physical affordance and social experience Vis (2013a) establishes that the boundaries by which the physically constructed composition of the built environment emerges would merit a further formal ontology. Vis (2013b) discusses how a theoretical conceptualisation of encountering, physically transforming, and inhabiting a built-up social and spatial world as a constitutive process in combination with a rigorous conceptualisation of the empirical information conveyed in configurative spatial layout data on the built environment can lead to an ontology of boundary types. These so-called boundary line types (BLT) (Table 1) are based on an abstract intellectual understanding of the differentiation between subdivisions and by their very definition simultaneously ensure their empirical identification in each specific urban situation. Vis (forthcoming) finalises and illustrates the formal definitions of these BLTs in expectation of a wider application and operationalisation in comparative urban built environment mapping processes. Within this appropriately themed volume, this chapter contextualises the method and discusses how to put BLT mapping into practice on the basis of archaeologically and historically acquired plans of urban built environments. The spatial data resulting from this practice enables a visual redescription of urban built environments on the basis of a rudimentary understanding of the socio-spatially affording characteristics of its physical presence, while the data structure is suitable for an array of spatial analyses in the future. This chapter will discuss how these methodological processes operate, but will not present the full analyses of any case study.

[1] Affordance is a concept derived from the work of psychologist Gibson (1979) on perception, which entails the quality of things and the environment which allows individuals all possibilities to act upon them. Here affordance is used in conjunction with the affective qualities of the experience of encountering things and/or the environment.

Table 1 This represents an abridged version of the full definitions and description of the boundary line type (see Vis forthcoming), which has been adapted from Vis (2013b)

	Boundary line type (BLT)	Empirically identifiable principle	Social relation to interaction opportunities	Exemplary indication (for western or globalised cities)
1	Closing boundaries	Operates on the basis of seclusion of a continuous spatial arrangement from the surrounding configuration with the material property that the boundary can be closed off towards its outside, thus making it a dominant. It is also a solid (i.e. no internal arrangement of outlines)	Interaction opportunities are quite stringently internalised as distinct from the outside, though there is a mutual (in)direct orientation between the solid dominant and the surrounding configuration	These boundaries typically circumscribe buildings of any sort or size
2	Facing boundaries	Operates on the principle of the orientation for soliciting interaction from the surrounding configuration	Is the site of solicitation of interaction with a dominant	These boundaries represent the doorways or entrance ways into a building
3	Associative boundaries	Operates on the basis of dependence on a single dominant. It is directly associated with and, in a conjunction including possible other (in)directly associated boundaries, with which it forms an adjoining configurative complex	Interaction opportunities are mediated between the openness of the surrounding configuration and the related dominant	These boundaries are typically associated with gardens or any plots and surfaces belonging to a specific building
4	Extended facing boundaries	Operates on the principle of orientation in an uninterrupted connection to a facing boundary by dependence on any boundary associated with a dominant	Is the site of indirect solicitation of interaction with a dominant, proceeding is no necessity	These boundaries are typically associated with garden gates or courtyard entrances, etc.
5	Directing boundaries	Operates on the basis that it directs interaction along opportunities for further boundary crossings in parallels	Interaction opportunities are directed along the boundary crossings that constitute its sides, connecting all sorts of bounded spaces	These boundaries are associated with the street network, access and pathways

(continued)

Table 1 (continued)

	Boundary line type (BLT)	Empirically identifiable principle	Social relation to interaction opportunities	Exemplary indication (for western or globalised cities)
6	Disclosing boundaries	Operates on the basis of guiding interaction towards opportunities for further boundary crossings in multiple directions rather than a single particular direction with necessary (in)direct connections to solid dominants	Interaction opportunities are freely organised, yet directed in multiple directions which in several cases will eventually lead to soliciting interaction with solid dominants	These boundaries are associated with square-like spaces in well integrated urban situations with several associated buildings
7	Enclosing boundaries	Operates on the basis of seclusion from the surrounding configuration with the material property that the boundary can be closed off towards its outside, making it a dominant while containing solid dominants	Interaction opportunities are restricted by solicitation between the openness of the integration within the boundary configuration and the configuration with solid dominants that it circumscribes	These boundaries are typically associated with city walls and gated communities
8	Mutual boundaries	Operates on the principle that it is simultaneously associated with or encompassing a distinct subset of several solid dominants with which it forms a configurative complex	Interaction opportunities are indirectly directed to several solid dominants and mediated between the openness of thoroughfare	These boundaries are associated with a specific group of buildings without any preference as to which it provides access such as shared porches, cul-de-sacs and communal space in gated communities
9	Opening boundaries	Operates on the principle that it creates open, accessible connections towards its outside, while being an integrated part of the configuration	Interaction opportunities are freely organised, with no prerequisites for boundary contexts and the possibility of thoroughfare	These boundaries can be described as park-like spaces, e.g. garden plots, urban fallow, parking surfaces
10	Neutral boundaries	Operates on the principle of neutrality, which results from ambiguity and the absence of singular associations, can occur in virtually any context	Due to the absence of a unambiguous relation to a residing socio-spatial system, crossing the boundary creates no difference from the surrounding non-dominant configuration	These boundaries tend to be the left over areas in less optimally used built environment configurations and also some delimited functional areas connected to streets (e.g. electricity supply)

(continued)

Table 1 (continued)

	Boundary line type (BLT)	Empirically identifiable principle	Social relation to interaction opportunities	Exemplary indication (for western or globalised cities)
11	Man-made boundaries of unoccupiability	Operates on the basis of negativity, can occur in most contexts	Negativity means there is no residing socio-spatial system, in this case because an area cannot be occupied by human beings	Structures that create a unoccupiable surface area, such as ponds, canals, architectural talus, narrow gaps, etc.
12	Not man-made boundaries of unoccupiability	Operates on the basis of negativity, can occur in most contexts	Negativity means there is no residing socio-spatial system, in this case because an area cannot be occupied by human beings	Steep slopes, natural bodies of water, etc., which are contained in the built environment
13	Not man-made negative boundaries	Operates on the basis of negativity, can occur in most contexts	Negativity means there is no residing socio-spatial system, in this case because it marks the end of the built environment	'Nature': wild or not fully cultivated areas
–	Virtual boundaries	Sites of distinction afforded by extant physical distinctions, human beings would have understood and/or experienced to be a crossing from subdivision into subdivision without clear material markers imposed onto the surface	Can in principle be part of any BLT that is not closable or negatively defined	Locations of crossings from space to space are in principle unimpeded and predominantly unmarked such as openings in dry stone walls circumscribing fields, or a cul-de-sac connecting to a street with similar surface

As mentioned previously, BLT mapping is intended to work as a comparative method and therefore should be applicable to the full breadth of urban built environments that societies have produced in the past and present. This is not to say that any comparison is inherently useful, nor does it preclude the possibility of even the most random comparison leading to a more profound understanding of inhabiting a specific built environment. Simultaneously, as the product of a high level social theoretical treatise, despite directly informing the empirical reality of urban data (cf. Smith 2011a), the understanding generated will be limited by an essentially human level of generalisation tied to its constitutive perspective on emplaced social and environmental interaction (Vis 2013a, b). The BLTs "operate firstly as distance setting, creating a personal territory; secondly choosing activities and project participation, negotiating the abilities and constraints; and thirdly the adherence to context, depending on the constitution and perception of (aggregate) entities. These spheres of significance intrinsically combine the social and spatial." (Vis 2013a: 26) The BLT concepts are a critically realist outcome of iterative abstraction (see Sayer 1981, 2000; Yeung 1997; Wallace 2011), which acts as a fundamentally human and social 'underlabourer' (cf. Pratt 1995) to undertake more detailed socio-cultural and historical investigations into the contexts of the inhabited built environment. Mapping BLTs therefore amounts to mapping the socio-spatial relations immediately afforded by and encountered as occurring in the physicality of the built environment: this process emphasises degrees of spatial dependence rather than detached spatially independent social activities (see Sayer 2000).

Predominantly, this chapter will make explicit the socio-spatially analytical mapping method in historical context arising from these thoughts, which retains the researcher in charge over the digital technology used (cf. Griffiths 2013). Therefore much space will be dedicated to the way the source data is prepared and the interpretive decisions made that allow the spatial data captured in GIS to preserve immediate access to the largely intellectually and empathically qualitative theory, which enables their individual identification. In doing so, GIS initially becomes an exploratory tool in which formally fixed conceptual data informs further insights into and precise contextualisations of the socio-spatial significance of inhabited built environments instead of a technological 'black box' (cf. Griffiths 2013; Lilley 2012).

2 Context of the Method

From the vantage point of comparative methods, it no longer matters that the settled world has known wildly differently shaped urban landscapes. Consequently, within this paper the mapping of BLTs will be discussed in relation to two very distinct test cases. These test cases, Chunchucmil (Classic Maya lowland, Yucatan, Mexico) and Winchester (1550-present, UK), have been selected on the basis of the basic quality of the data available and their arbitrarily contrasting general configurative characteristics. The Maya culture area is known to have been the home to a low-density (tropical) agrarian urban tradition (see Isendahl and Smith 2013; Fletcher 2009; Graham 1999), while Winchester is a typical historical English (and western European) example of a densely settled urban landscape based on a persistent medieval pattern with Iron Age and Roman antecedents (see Conzen 1960). High-density urban settlements have become the default of the globalised western city, but low-density urbanism was until recently (pre-European, 19th century) also a tradition widely found in Africa (Smith 2011b).

The method of mapping boundary line types does not stand isolated in its efforts to capture social and spatial, or historical and spatial, information on the basis of built environment layout ground plans.

2.1 Urban Morphology and Town-Plan Analysis

Town plan analysis took a firm hold in geographical practice on account of Conzenian urban morphology (Moudon 1997; Whitehand 2007). Its combination of streets, plots, uses and fabrics, was grounded specifically in spatial and historical rather than immediately social interests. Town plan analysis aimed to reveal the building history of the shape of a town based on historically and spatially coherent plan units, which are somewhat subjectively identified by the researcher (see Conzen 1960, 1968, 1981; Whitehand 1981; Lilley 2000; Conzen 2004). It thrives on large bodies of social and economic historical sources and a degree of intuition, which are an intrinsic part of the spatial presentation of the research outcomes in its resultant urban mappings. It structurally connects historical context to the processes that shape urban space to create a townscape created out of the town's plan, building fabric and land utility. The town plan in turn is composed of three elements: the street pattern, the plot and aggregate blocks, and the buildings and their block plans (Conzen 1968). Yet, ultimately the analytical unit of urban morphology, the plan unit, is a relatively coarse spatial reference, which cannot reconstruct a town's precursory phases in great detail.

Not all historically conjectural or regressive sequence mappings of urban plans result from urban morphology, but the research practices which Conzenian urban morphology sprang from (see Lilley 2000), seem to have paved the way for the more general acceptance of recent historical practice. The work of Keene (1985) on medieval Winchester, drawn upon in this chapter, Keene and Harding (1987), Dean (2012a, b), and Bisschops (2012), amongst others, demonstrates degrees of similarity with urban morphological regressive mapping practice, as most specifically morphologically practiced by Lilley (2000, 2011a) and Lilley et al. (2007). As Bisschops (2012) points out, his practice uses both historically intensive regressive mapping and the 'cross-section method' inspired by Keene (1985). This approach allows series of properties to be mapped with reasonable accuracy, while anchoring incidental properties or buildings within each sequence of properties (cf. the notion of plot series, Conzen 1960).

The aim in this type of work tends not to be the historical explanation of the origins of the formation of the urban form and building fabric of the town as in traditional urban morphology, but to reconstruct the town plan for a specified moment or period in the past. This is then used to geographically locate and position (social) historical sources to the spatial shapes and references created, now greatly aided by the data management technology within GIS. Urban morphology, in part, may help create a skeletal plan for earlier phases of a city often on the basis of the first accurate urban plans from the 19th century. More advanced, intensive historical and archaeological methods (Lilley 2011a; Bisschops 2012; Dean 2012a, b) of working backwards, processes of careful cartographic reconstruction, are needed to approach approximately complete and reasonably accurate snapshots representing the historical town. This practice has only been developed within western European contexts. In that area the reconstructed plans resultant from regressive sequence mapping of urban layouts form a methodological prerequisite for the analytical socio-spatial mapping method presented here, especially where ongoing cities inhibit large extents of archaeological mapping.

2.2 *Space Syntax*

Perhaps the best known analytical approach to ground plan configurations is the family of theory and tools going under the header of 'space syntax' (Hillier and Hanson 1984; Hillier 2007; Bafna 2003), which is increasingly claiming its place in archaeology (Thaler 2005; Van Nes in preparation; Morton et al. 2012a, b, this volume; Fisher 2009; Stöger 2011; Paliou and Knight 2010). There are several important similarities between the methodology presented here and urban space syntax, which are not purely coincidental or to be ascribed to using a significantly concurrent dataset. Both emphasise topological information and both seek a social understanding of certain structural properties of the built environment's configuration. However, space syntax privileges the open spaces left within the built-up morphology, which logically leads to movement across space as the most successful area of space syntactic research: navigational (pedestrian) movement in particular (Bafna 2003; Morton et al. 2012b, this volume; Craane 2009).

Space syntax tools, on an urban level,[2] calculate probabilistically or rule based spatial patterns across its 'open' space, afforded by the configuration. It is this space syntactic pattern, an abstraction derived from a simplification into convex spaces rather than the original empirical reality of the shape of the configuration that is analysed to produce an array of values (see Batty and Rana 2004). Space

[2] Space syntax incorporates a family of analyses, which also includes a strong branch in interior spaces, based on graph analysis (Hillier and Hanson 1984). Incidentally, in archaeology syntactic analyses of interior space are arguably yet more wide-spread than urban analyses (see Fisher 2009).

syntax in general is often strongly linked with arguments in spatial cognition (see Bafna 2003) and omits a clear incorporation of phenomenological or individually experiential aspects of built environment configurations (see Griffiths and Quick 2005), except for the development of space syntactic viewsheds (isovists) (see Franz and Wiener 2008). Moreover, the theoretical underpinnings of space syntax do not intrinsically include a historical or constitutive grounding. These are important reasons for Griffiths (2005, 2011), to urge a careful and critical use of space syntax only in historical research.

It is worth bearing in mind that space syntax was born out of attempts to understand "the influence of architectural design on the existing social problems in many housing estates that were being built in the UK" (Pinho and Oliviera 2009: 110) and therefore to better inform urban design practice (Hillier and Hanson 1984). In general space syntactic analysis comments on the probability of the built environment configuration affording lively or less lively streets, which is often connected to economic viability (e.g. Chiaradia et al. 2008; Narvaez et al. 2012; Griffiths et al. 2010; Valente 2012). The probability of liveliness is associated with the cognitive readability or intelligibility of space for way-finding argued to be tied to global and local integration measures (Van Nes in preparation). The separate sides of the street in some analyses have an abstract presence restored through the measures of the 'constitutedness' of streets, expressing the quantification of buildings opening out onto the street (e.g. Van Nes and Lopez 2007; Palaiologou and Vaughan 2012). The prevalence of the street as a clearly defined space is also found in urban morphology and in both cases limits the potential comparative application of the methods. Maya cities, for example, are known to have few clear streets (Magnoni et al. 2012), while there exists much open space that is organised differently.

As will be demonstrated, the BLT mapping presented here, adheres to the original physical morphology of the configuration, maintaining much of the material spatial information. It is based on constitutive theory and is directly experiential in its use of Gibsonian affordance (Vis 2013a; Vis 2009; see Gibson 1979) rather than using it for cognitive assertions. It seeks to elicit the features that afford opportunities for further socio-spatial development and use through time by inhabiting it. BLT mapping socio-spatially conditions and contextualises the encounters or actions that could take place within each emergent location. Although the boundary method necessarily is inclusive of interconnectivity and affordance of movement across the built environment, it does not make any probabilistic claims about that movement itself. Instead, the potential for movement or any other use becomes qualitatively characterised and its socio-spatial position within the configuration contextualised in formal socio-spatial descriptions of each space. This may narrow down likely functions within space (various actions are possible in the same space, cf. Sayer's (2000) spatial independence), but does not express the probability of something occurring within a specific space. After all, BLT mapping is limited to the conduciveness of the physical characteristics of the distinctions making up the built environment and not, like space syntax, to a pattern of its linked open spaces.

2.3 Historical Objects of Urban Fabric

A final methodological development of relevance here is championed by Lefebvre and others (Lefebvre et al. 2008; Lefebvre 2009; Rodier et al. 2009; Lefebvre 2012). They have been building a conceptual ordering of the urban fabric that emphasises temporal dynamics and function (the underlying theoretical model goes under the name: OH_FET) for storing and analysing urban archaeological information. This practice is based on a conceptual model hierarchically composed of simple and (aggregate) complex objects elucidating the intricate formation and use of architectural complexes in an urban setting (Lefebvre et al. 2008; Lefebvre 2012). It focuses on eliciting the historical rhythms of built space in development. Following Galinié et al. (2004), the method is based on accepting the assertion that any understanding of the dynamics of urban fabric over time necessitates the conceptualisation of a constituent object of the urban fabric which compiles all knowledge about its transformations. This gives rise to the 'historical object' an initial interpretation meeting three fundamental criteria: (1) location and surface area (where is it?); (2) date, duration and chronology (when did it exist?); (3) function, social use, or an interpretation (what is it?) (Lefebvre et al. 2008; Lefebvre 2009; Rodier et al. 2009). Lefebvre (2009) explains that any modification of these three criteria causes the disappearance and creation of a new historical object or interpretation. Theoretically this is a logical consequence of the aggregate complexity of historical objects. It should be noted that this endeavour includes more information sets than the socio-spatial significance pursued by BLT mapping, together with the ultimate goal of understanding the dynamics of urban formation. To that end, detailed temporal information is the driving force for generating analytical spatial units, while it remains unclear what the meaning of these features is.

Lynch (1981) insightfully remarked that many building notions conflate the space it represents with the type of use associated with it (e.g. a church). Inevitably, the social use types found in Lefebvre (2009) and Rodier et al. (2009) are largely examples of such notions. They combine established cultural interpretations, which exist only in relation to specific cases and could additionally be dependent on historical periods. This will often preclude broad comparative application. Another problem is the desire to treat temporal intricacy on a similar level as spatial complexity, as this would require equal information across the whole urban space for each unique moment of (spatial or functional) transformation. This introduces a laudable pursuit of a research aim privileging temporal dynamics over the more conventional time-slice or snapshot approach. Yet, this concern runs the risk of overlooking that equal historical information is required across the whole of the research area at each separate instance of change within it, which is rarely the case and thus introduces conceptual anachronisms or conceals inevitable extrapolations. In contrast, time-slices are necessarily coarser and more conjectural, but in their atomic assumption[3] of the (historical or reconstructed) urban plan (see Vis 2013b), they are better suited for spatial analysis. The discrepancy between methods is caused by disaggregating complex historical objects into more temporally specific features. Logically such an approach is naturally better suited for intensive studies of smaller urban areas, as demonstrated by Lefebvre (2012).

The boundary method, however, does not emphasise temporality, but rather the social reality of physical space at certain moments. Due to the method's comparative aims, categorising particular functions or uses of space is refrained from. As reconstructed time-slice plans in the historical dimension are accepted, it does not fully depend on equal historical information. Instead, a moment is chosen for which the best consistent information is available or for which conjectures are justifiable across the area as a whole (cf. Lilley, *Mapping Medieval Chester*). The BLTs focus on the spatial embeddedness of occupying (thus using) spaces within the built environment at any such moment. In contrast to Lefebvre and consorts, with this approach, temporal details are lost. BLT mapping thus privileges the general social dynamics of physical space over those of time, although its theoretical basis ensures a historically constitutive understanding of the conditions of living in the built environment. The aims of urban morphological approaches, space syntax and OH_FET adaptations all differ quite specifically, while maintaining some degrees of similar reasoning and practical compatibility with the boundary method as outlines in this paper.

3 Mapping Boundary Line Types

Mapping boundary line types (BLTs) is an analytical and interpretive process, which requires high quality source material. As mentioned before, within this paper the process will be demonstrated by using two test cases: Chunchucmil and Winchester. Not only are the cities themselves enormously different, the source material available also represents two distinct disciplinary products. The test case in Chunchucmil relies fully on archaeological material from over a decade of intensive surface surveying on site (Hutson et al. 2008), while the Winchester test case relies on material meticulously compiled and integrated from extensive medieval historical and cartographical research (from here: 1550s) (Keene 1985), the first edition large 1:500 scale, OS survey city plan of 1872 (from here: OS 1872), and the current large scale OS MasterMap (from here: MM). Their selection thus serves the double purpose of explaining the variants of the BLT mapping practice depending on historical and archaeological source material, while at the same time explicating the comparative potential of the method.

[3] Assuming a time-slice is atomic explicates its momentary indivisible nature as a whole. A time-slice is an abstract entirety which is immediate and inseparable: no time passes, everything occurs at once.

All maps are necessarily selective representations of spatial information and never an incontestable truth (Wood 1992; Monmonier 1996; MacEachren 2004; Lilley 2011a, 2012; Gauthiez and Zeller this volume, Beisaw and Gibb 2013). In the method at hand GIS software is used to create the spatial data (maps) appropriate for the current purpose (city ground-level outline base plans). The base plans upon which BLT mapping can be applied are interpretively derived from original mapping sources. Chunchucmil's source material concerns an archaeological plan, prepared in Adobe Illustrator software. Following archaeological practice, this depicts all that was observed and identified as archaeological features. Winchester's source material concerns a historically reconstructed and conjectured plan of the property boundaries in the 1550s, a scan of the historical sheets of OS 1872, and MM: the current Ordnance Survey large scale standard. The purposeful character of each of these maps is distinct, according to the differing aims, selections and technologies of the mapmakers. They need to be especially prepared not only to be imported into a GIS environment (here: ESRI ArcGIS), but to establish an equal and comparative foundation of the selected spatial data (see Vis 2013b).

The idea of a (ground-level) base plan as an initial stage of practice features in various kinds of urban analysis. Urban morphologists would prepare a base plan on the basis of a good quality 19th century map (see Conzen 1960; Lilley 2000), and in space syntax the base plan drawing typically separates open (street) space from blocks of buildings (Hillier and Hanson 1984; Van Nes in preparation). For BLT mapping the base plan follows from an interpretive process of decision making on the nature of the physical distinctions of the constructed subdivisions that shape the built environment's composition (Vis forthcoming). The subdivisions are only represented by their outlines (i.e. their empirically recognisable boundaries) and not by the internal and functional designs of the emergent areas (Vis 2013b, forthcoming). Inevitably, this process in part relies on professional judgments on what to include and what to exclude, as elaborated in the examples below.

This is the first time BLT mapping of the urban built environment has been put into practice. What is reported on is a digital adaptation of the process. However, during the initial trials of the mapping process it was first tested with pens and semi-transparent paper (i.e. manual GIS layers), to see if and how the source material would be suitable for this work. The presented method is aimed primarily at what are known as 'legacy datasets': basically extant maps and spatial data produced in earlier projects and/or for other purposes. Being able to work with legacy data greatly increases the wide application and flexibility of this method. Yet, one is also confronted with issues concerning the nature of the spatial data that may not be immediately appropriate for all aspects of this method.

The workflow could be summarised in the following steps, further discussed below:

1. Acquiring or assembling and converting the source material (typically legacy spatial data and/or maps);
2. Creating equivalent spatial information by mapping (tracing) the outlines of main features contained in the source material;

3. Case specific conjecturing to resolve data gaps and revising the resultant outline base plan;
4. Remapping the base plan by identifying the BLTs.

4 Chunchucmil Site Plan

The archaeological plan of Chunchucmil was originally acquired in digital format from Scott Hutson, University of Kentucky, who directed and completed the mapping project on Chunchucmil over the past few years, taking over from Bruce Dahlin. Frequent contact with Scott was invaluable to using the still unpublished map (Hutson et al. in preparation), which necessarily depicts archaeological interpretations of the features mapped, and gets to grips with the empirical situation encountered on-site which the map does not convey. Hutson et al. (2008) demonstrate that dateable artefact assemblages show a relatively equal period occupation across the full surface of the site (5th–6th century, Classic). There is a central barricaded portion that indicates Terminal Classic and later reuse of the monumental core (Dahlin 2000), with little activity elsewhere. Therefore, knowledge about the palimpsests and possible anachronisms of the remains observed is limited. Where a mapped feature is cut by another, thus making the former largely obsolete, here only the latter is taken into account. The resultant reading of the site's plan should be seen as a rough approximation of a final stage of maximum occupation throughout the site.

Chunchucmil's plan is prepared as an 'ai' (Adobe Illustrator) extension, which is dedicated to adaptable graphic presentation rather than (geospatial) data storage. Unfortunately, this file format cannot be directly imported in ArcGIS. A laborious conversion process for rescuing legacy Illustrator data, set out in Wunderlich and Hatcher (2009), was roughly followed. This involves, amongst other steps, manually separating out layers of disparate digital information (e.g. lines and text) and automatically densifying the distribution of anchor points (vertices) along lines to preserve software generated shapes. The route to ArcGIS proprietary shape file ('.shp') format then requires AutoCAD file formats (a hereditary '.dxf' exchange format (2000/LT2000) was used, which proved more stable than newer versions). Though this format should suffice for direct import into ArcGIS or freeware programmes like QGIS, such conversions produced compromising results. It was found that MapInfo's Universal Translator tools get reliable results (first to proprietary '.tab', then from '.tab' to '.shp'), which lead to directly usable shape files. Both a raster image—for visual clarity aiding the reading of the new vector data, especially needed since Illustrator annotations did not make it through the conversion successfully—and the shape files of the original plan were imported as GIS layers.

In order to inspect the converted data quality visually, the shape files and raster image ('.tiff') need to be geographically related to each other. This is achieved by projecting and georeferencing the data. Although several GPS points were taken across Chunchucmil (Hutson, personal communication), the inherent error margins associated with such points, would result in unequal placement and georectification between the data layers. Therefore only the site's centre GPS point was used. This point forms the centre of a derived 250 × 250 m grid over the core of the site, for which the coordinates can be calculated. When separating the layers in Illustrator, this grid was transferred into each layer. In doing so, the raster image could be precisely georeferenced according to the on-site grid system. Subsequently, the calculated coordinates allowed exact spatial adjustment (placement) of each vector data layer on top of the image. Then it is possible to inspect visually, per individual detail, whether the conversion is complete and if not, to reveal any missing data.

4.1 Chunchucmil Base Plan

From now on the work becomes dedicated to selecting the basic spatial data that is conceptually required for the BLTs to be identified in the next stage. The theoretical conceptualisation of filled space and the inhabited built environment (Vis 2013a, b) necessitates a reduction of information to the physically binding built boundaries, which also function as empirical counterparts to conceptual 'sites of difference' (cf. Abbott 1995). These physical distinctions separate spaces into discrete subdivisions (see Vis 2013b for the underlying progressive processing of spatial-physical information). Since the Chunchucmil plan now exists as vector data, the process of creating a base plan of outline features is predominantly limited to using the 'tracing' function in ArcGIS editor. In addition, regular editing tools were used for conjecturing any apparent gaps in information (see below).

Tracing the lines interpreted to be outlines revealed some issues with the data structure and topological integrity, most probably resulting from the initial drawing of the map in Illustrator.

Visually presentable figures would in minute detail reveal line constructions that were unsnapped or simply did not match the features contiguous geometry (Fig. 1). Tracing requires continuous lines. Despite these errors usually measuring no more than a few centimetres or even millimetres in real space, literal copying of the data would not produce usable outline data. This means that the process of tracing outlines also required additional editing for cleaning up and sometimes redrawing features, ensuring a proper structure for the outlines represented by a single polyline. At times, features of different layers representing two classifications of information (e.g. architecture and boundary walls) would come conspicuously close to connecting, yet virtually never would these features truly connect. As a rule of thumb, detached mapped features of equal or differently interpreted classifications would be connected within the GIS layer (polyline feature class) used to prepare the base plan if approximately under 80 cm of width. Any analogous gaps of larger width or conspicuously positioned would be connected in a separate conjectural layer.

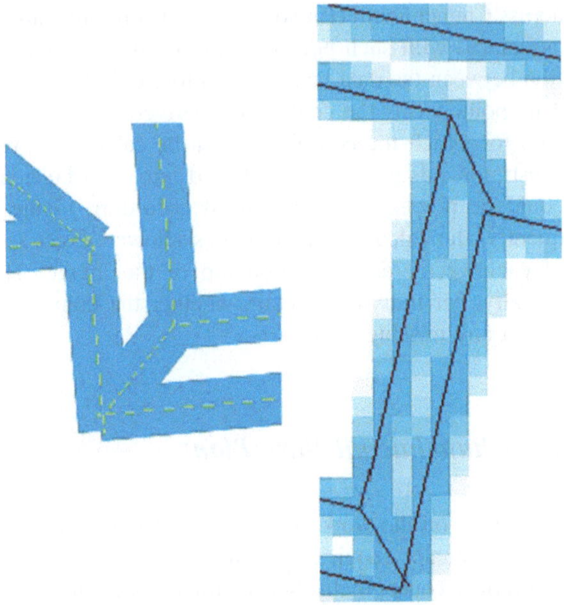

Fig. 1 Illustrator data (*left*) and converted ArcGIS shape files on top of raster image (*right*). (Image appears courtesy of the Pakbeh Regional Economy Program with help from S. Hutson.)

Conjectures were also used more progressively after having traced all originally observed features deemed to represent outlines. As expected in even the best archaeological information, the fragmentary nature of material remains means that various, clearly unintended, gaps in the data persist. The theoretical foundation requires all the integral subdivisions of the built environment to be present as a base layer of information, so such remaining gaps needed to be critically filled with harsher interpretive conjectures. As to date no other example of a Classic Maya site is known to manifest such a constellation of elaborate house groupings, pathways and boundary walls in the areas outside of the monumental centre (see Hutson et al. 2008; Magnoni et al. 2012) conjecturing analogically is largely out of reach. Therefore a rather bold approach to conjecturing was decided on. Fragmented buildings would be finished continuing the shapes suggested in the observed remains. Fragmented boundary walls are completed exclusively with straight lines (without crossing any other tracings), directly connecting two ends of mapped lines of the same class or onto another feature, using parallel and perpendicular alignments (see Fig. 2). Especially in the highly non-geometrical urban form of Chunchucmil, straight lines emphasise that these conjectures are not intended to represent informed reconstructions of the actual features' shapes. They rather complete the spatial data by reconstructing close approximations of the likely topological relations that would have existed between subdivisions. This unavoidably compromises the morphological integrity of the data, but to suggest actual knowledge of the features amounts to over-interpretation. Conjectured information can always be retrieved as this is kept as a separate dataset (see Fig. 2).

Fig. 2 Outline features (*in grey*) and minor and coarse conjectures (*pink*). (Image prepared from original data courtesy of the Pakbeh Regional Economy Program with help from S. Hutson.)

It should be understood that these crude conjectures are a requirement of the conceptualisations behind the mapping practice being discussed. It is not suggested here as general archaeological practice and is not a necessity for each form of interpretation on the basis of the plan (as demonstrated in Magnoni et al. 2012). Nevertheless, for any future topological spatial analyses, the conjectures are necessary. In Chunchucmil's case, the complementary conjectures went through three iterations. The initial phase concerned the coarse connecting up of features on screen. Then a revision was carried out based on the principle that directly or indirectly all spaces within an urban environment must be accessible. In other words, how is one able to traverse the site respecting the actual physical barriers mapped? This intuitive patterning was carried out on a high-resolution printout of the map with a semi-transparent overlay. The possible lines of movement and flow across the site were drawn. So the conjectures were adjusted to enable or better facilitate necessary traversing. A final revision is a side effect of the actual process of BLT identification, which would point out any subdivisions where a specific discrete outline was commonsensically expected, but absent (Fig. 2).[4]

Since the Chunchucmil's base plan aims to convey the most basic physically constructed subdivisions, archaeological artefacts, stelae and quarries are generally

[4] It is important to note that all conjectures can be retrieved from direct comparison with the traced data of the original plan. It is likely that more means become available to improve on the conjecturing (or reconstruction) process in an informed way if additional research is carried out, thus further giving reasons to revise and adjust insights derived from the current iteration.

not included, unless a quarry seemed suggestive of possible incorporation as part of a (conceptual) boundary. Querns or *metates* within gaps in walls are taken as an indication of a probable passage way because the arduous task of grinding in all probablility had a social element to it (Hutson, personal communication). Bedrock, however, is included as these outcrops of the natural soil would have impeded thoroughfare and are often incorporated in boundary walls and even minor architecture. One should always be mindful of the possibility that consciously or subconsciously, readily perceived or concealed, subjective patterns could emerge from the underlying decisions and rules of thumb regarding outlines and conjecturing.

While the base plan consists of tracings of originally mapped outlines and conjectures, there is an intellectual aspect to boundaries that is not immediately represented. This is the 'virtual boundary' (see Table 1). A fiat boundary is understood to be present, but is not physically there (see Smith and Varzi 1997; Smith 2001; cf. Vis 2013b for a treatise on how fiat and bona fide boundaries are used in understanding spatial data). For Chunchucmil's boundary walls, it would be a logical expectation that there were (possibly closable) openings in them which allowed people to access the areas they circumscribe. However, in order to create discrete outlines as well as accounting for possible data gaps, these openings would have been connected up by conjectured lines. There is no way to distinguish on the basis of the mapped material remains whether any opening was intended, destructed, deteriorated, caused by decayed perishables (e.g. incorporated cacti), or removed (by animals or humans) after abandonment, etc. (Hutson, personal communication; Becker 2001; Demarest 1997). At the same time there is no real necessity for each wall or distinction to complete a circumscription contiguously. Indeed, in Chunchucmil several platforms tying building groups together would gradually descend into the ground, creating a ramp for probable unimpeded access (Hutson, personal communication). Virtual boundaries, though not themselves a BLT (Table 1), are thus used to mark-up situations in which it is likely that missing physical information would not have detracted people from contextually recognising the spatial distinction at that location. This enables further discrete subdivisions to be mapped (Fig. 3). Note, however, that although virtual boundaries would note places of unimpeded access, entrances (represented in Table 1 by two separate BLTs) do not only operate on virtuality, nor do virtual boundaries only occur on the basis of conjectures.

Although an outline base plan is necessary for BLT identification, it cannot be assumed that the prepared base plan will contain all the lines as present after the full BLT identification process. Identifying BLTs, although bound to adhere to their definitions (Vis forthcoming), is still an interpretive practice on top of the base plan. While 'dangling' physical distinctions do not immediately cause a discrete subdivision, in preparing the base plan these would still be included, as final informed selections follow from the BLT identifications. In some cases these dangles originally traced from the site plan may be decided to not support a fully-fledged subdivision. They then become obsolete, as feature specific to the internal and/or specifically functional arrangement within subdivisions. Furthermore, it can be decided that virtual boundaries should be adjusted to support a commonsensical or more basic boundary line structure. Nor does a base plan necessarily include all the virtuals, as the most regular virtuals result from applying the BLT definitions themselves. In other words, a full base plan can actually only be derived after all BLT identifications have been carried out. All unique BLT identification polylines form a full and selective outline base plan (Figs. 3, 4).

In Maya settlement archaeology, Becker (2001) strongly argues for the existence of perishable boundary walls and (formal) hedges. Although predominantly stone and rubble were used in Chunchucmil, it is entirely possible that what we no longer see, could have perished. Much of what may have perished could have made internal divisions of activity areas within the bounded areas of boundary walls (*albarradas*) (Becker 2001; Manzanilla and Barba 1990; Hutson et al. 2007; Hutson et al. 2008). Similarly, various structures within groups of buildings could have been perishable (Becker 2001; Magnoni et al. 2012). Here, if nothing at all was mapped, no additional conjectures are invented. However, there are ambiguities such as the 'screen walls' connecting structures in building groups in various Classic sites, mentioned by Becker (2001) and Tourtellot (1988) (see Fig. 4). These could easily be confused with remnants of communal platforms. In such cases the expert opinion of the mapmakers indicating where they identified a platform or group on the plan (by annotating it) is cautiously followed. Although Magnoni et al. (2012) offer a population estimate which is partly based on the count of residential structures (following the method of Rice and Culbert 1990), for the purposes here the different structures which will be part of the building

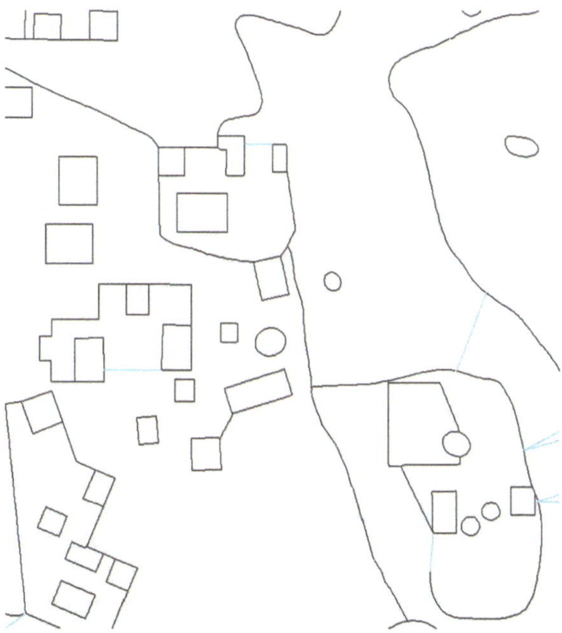

Fig. 3 Chunchucmil's outline base plan (*grey*) with several virtual boundaries (*light blue*). The virtuals on the right hand side would not have originally been included in the base plan. (Image prepared from original data courtesy of the Pakbeh Regional Economy Program with help from S. Hutson.)

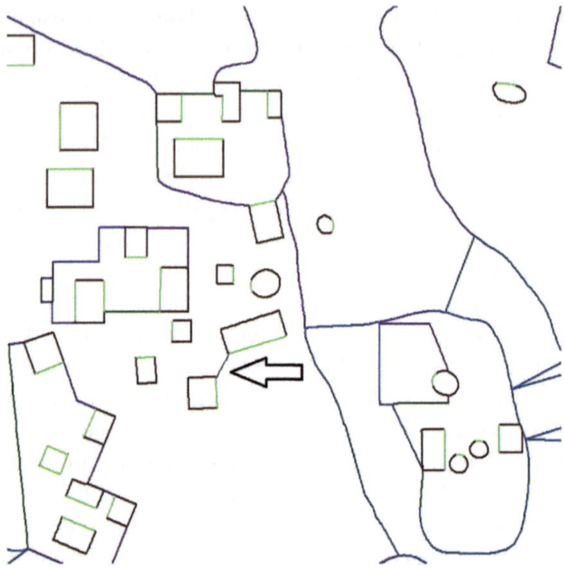

Fig. 4 The *arrow* indicates a line which was previously part of the base plan and could not successfully be completed into a full subdivision. Sometimes buildings could have a wall between them because of the wind (*screen wall*), thus a functional feature not fully separating anything socially. (Image prepared from original data courtesy of the Pakbeh Regional Economy Program with help from S. Hutson.)

group are not considered per function. Becker (2001) gives a good overview of what functional structures could entail, but also how few of them are systematically identifiable. Various fragments of albarradas can also be found 'dangling' inside house-group-lots. These dangling lines may well form part of internal arrangements in concordance with the activity areas and perishable boundaries. With this in mind, these dangles can be excluded from the base plan. The activities together are part of the same, larger, socio-spatial system mapped by the outline.

4.2 Chunchucmil BLT Identifications

Despite using digital tools, creating the base plan is mainly a manual process within the GIS environment. BLT identifications in turn are based on manual tracings of each instance, individually and entirely; a BLT is successfully recognised within the base plan. For the base plan to be effectively supportive of this work, one needs to ensure topological integrity throughout the dataset.

ArcGIS has developed the 'topology toolbar' for the purpose of ensuring topological integrity. For this to work it is most convenient to merge any separate layers composing the base plan into one comprehensive dataset. Within ArcCatalog any feature class (layer) in a geodatabase can be subjected to a topological rule set. When validated, any errors can be inspected and corrected within ArcMap. The topology rules first let one set a cluster tolerance, simplifying the data structure (any features below a measured threshold are regarded the same), and further allow one to make sure the data does not contain any unintended dangling lines, unsnapped vertices or nodes, intersections, unwanted duplication or coverages, unconnected polylines, etc. Despite functional limitations[5] this semi-automated process both speeds up subsequent work and improves the immediate quality of derivative data. Any unfinished but suggestive subdivisions kept in the base plan as well as any 'edge effects' (unfinished subdivisions truncated by the test area selection) can be marked as exceptions in the correcting process. The cluster tolerance should be a measure commensurate with the precision achieved in the mapping resolution. For Chunchucmil 10 cm was selected, which might be smaller than the actual mapping resolution achieved on-site, but which would retain most of the interpreted shapes (e.g. curves) generated from the observations in Illustrator (see the above comment on densifying anchor points[6]).

As boundary mapping is based on outlines, all layers will always consist of polyline features. The paramount difference with the data structure of the outline base plan and each separate layer conveying a BLT is that each polyline feature in the BLT layers must convey a BLT identification in its entirety. In contrast, for the base plan the only truly important characteristic of the data structure is that, when maintaining topological integrity, all lines shaping the outlines are there. In other words, each separate feature in each BLT layer is a complete meaning carrying empirical identification of a material socio-spatial occurrence at the time represented by the plan.

The BLTs (Table 1) are numbered according to an order which is principally a logical order for carrying out the BLT identification process. As the ontological primacy of the solid dominants was established (Vis forthcoming), the first stage of BLT identification would see the researcher create a visualisation resembling a figure-ground plan, mapping out the built volumes from the open spaces (see Trancik 1986). This follows from Type 1, closing boundaries, creating a materially closable extraction from its surroundings (i.e. a complete structure or building). After identifying Type 1s, one would continue perusing the configurative relations to their surroundings and from this and the shape of the Type 1s, start identifying facing boundaries (entrances, Type 2). Taking in more of the surrounding configuration incrementally, Type 3s, associative boundaries, can be identified as bounded areas in direct association with a single Type 1. Type 3s could get a Type 4 (entrance, or extended facing boundary) in direct relation with a Type 2, to extend a mediated relationship of the Type 1 with its surroundings. If no Type 3

---

[5] It is also possible to set topological rules before mapping and check up on data created in a (semi) live way, whilst editing the data. This could be more efficient if most eventualities are known upfront. Likewise topological rule sets can be adjusted if it is found that the rules do not adhere to the intended logic. Unfortunately, topological rules appear unable to handle composite rules regarding more than one feature class (layer) at once, but can only run multiple questions simultaneously, each treating a single layer. Complementary coverages can be checked by using tools for selecting on location.

[6] Of course the simplification process is not 'intelligent'. Though accuracy to about 10 cm is maintained, at very large scales, some shapes might have become altered counter intuitively, e.g. right angles might have become slightly flattened and curves less smooth. Unwanted changes can be manually adjusted.

can be found, within most built environments a Type 5–9 could be expected (Fig. 5). Note, however, that while Type 1s need at least one Type 2 to socially partake in the configuration and avoid a negative definition (Types 11–13), beyond this there is no necessity for any specific BLT to occur in the direct vicinity. The definitions will allow the successful identification of each specific BLT that does occur as part of the outline base plan in which Type 1s and 2s are now defined.

The limited area in Fig. 5 demonstrates there is not a one-to-one relationship between a BLT identification and a line of the base plan. All segments of lines are identified as two or more BLTs (see Vis forthcoming). This reflects the binary character of any boundary, which can be approached from either one of its sides, thus affecting the role or significance it has for the person located in space.

From a purely intellectual point of view, the BLTs that do not circumscribe a subdivision but qualify the relation between two subdivisions, i.e. Type 2 and Type 4, including the virtual boundaries, should be drawn twice. Ultimately each

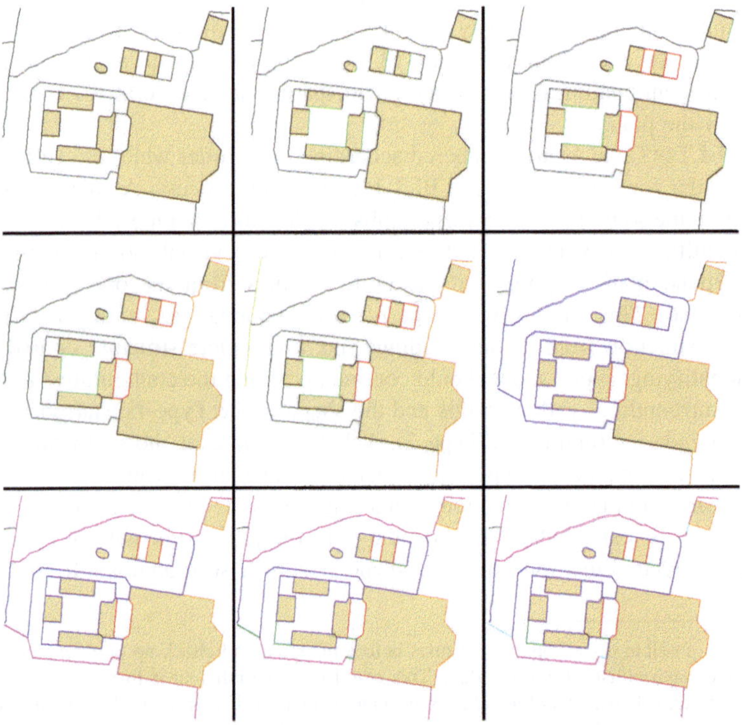

Fig. 5 The BLT identification process in nine stages on the basis of the Chunchucmil outline base plan (*black*). From left to right, in order of the BLT identification added: Type 1 (*brown*); Type 2 (*light green*); Type 3 (*red*); Type 6 (*orange*); Type 7 (*yellow*); Type 8 (*purple*); Type 9 (*pink*); Type 4 (*dark green*); virtual boundaries (*clear blue*). For visual clarity Type 1 has been converted into filled polygons. Note that BLTs that surround features get both an inner and outer identification. (Image prepared from original data courtesy of the Pakbeh Regional Economy Program with help from S. Hutson.)

entrance or virtual qualifier of a boundary is part of each colocated BLT (e.g. the reciprocity of any passage or doorway). This would essentially duplicate identifications at such places. However, for both visual and analytical purposes this is unnecessary in the spatial data. The length, construction and location of any such boundary could be recognised and utilised from the single identification of such boundary qualifiers.

If the data resulting from the BLT identifications should be used for computational spatial analyses or converted into other formats, then it is paramount that the BLT feature classes are checked against an appropriate topology rule set as discussed before. Fortunately, the polyline features are a flexible format, which can easily be processed into closed polygons, which would enable any investigation or visualisation on the basis of areas rather than boundary lines, while vice versa the intricacies of the BLT data structure cannot be automatically preserved. Computational spatial analyses are not part of this current paper, but are pursued in the continuation of this ongoing research.

4.3 Chunchucmil BLTs

Identifying BLTs effectively does rely on a series of informed professional judgments based on rules of thumb on how to read the plan. Entrances are especially tricky to positively and exclusively identify. The nature of the archaeological map is such that from the piles of rubble and debris from which the mappers were able to estimate structures, it would typically not be possible to also define their entrances. This leads to the use of an analogical rule of thumb by giving preference to buildings facing each other (directly and indirectly) as found throughout the Classic period in the Maya area in plaza and platform groups (see Becker 2001, 2004). Additional entrances may depend on any associated boundaries in their configurative complex (see Table 1). Alternative locations for entrances are identified when the spatial context displays a great measure of orientation elsewhere rather than in a facing fashion between buildings. Hutson (personal communication) suggested that small structures in the middle of plaza groups could have been entered from any side as they could have served as elevations to address audiences.

It could generally be assumed that platforms are accessible from all angles as they are low enough to mount without too much trouble. Similar to encountering a low fence, however, it would be logical that there are preferred places to access a platform. In many cases the platforms have been mapped as if they gradually descend into the ground (see also Fig. 5), as confirmed by a detailed excavation of a platform group (Hutson, Personal Communication), on one or more sides. Caution is necessary as a discontinuous platform outline could have a number of other causes besides those of intentional architectural construction. Subsiding platform sides in conspicuous locations within their context are taken as an indication for places to ascend onto or to access the platform. In instances where there is a full outline (possibly including a conjunction with albarradas), a wider opening between buildings or orientation towards the surrounding configuration is accepted as the indicator of an access way.

As a rule of thumb, albarradas are regarded to be physically impermeable, but similarly mitigated, due to heights lower than the human field of vision and the conspicuously fragmented and often virtual nature of their course, which occurs even in well preserved areas. Impermeability is thus mitigated by probably wide and/or multiple passages. Only albarradas mapped over a complete circumscribing course could become identified as enclosing boundaries (Type 7), which is also considered for rare high platforms, the outlines of which are often formed in conjunction with structures, with a probable formal entrance.

The parallel definition of Type 5 is applied in a very relative sense, sometimes including mirroring lines and contextually derived directionality. This means that two lines forming a relatively narrow (in context), but irregularly shaped corridor in a mutual linear orientation are likely to be defined as a type 5, broadly applying empirical parallelism. Type 5s running long contiguous courses are rare in Chunchucmil. The decision between a Type 5 or Type 9 is subjective to the degree that the researcher needs to judge when the general observed parallel structure is sufficiently lost.

Though Type 6 depends on opening out onto a few Type 1s, it is set apart from Type 8 because of its integration and sense of local centrality. It would seem that plaza and platform groups make good candidates for Type 6s, but usually their bounded area is spatially removed from the 'open' flows of traversing the site from anywhere within the spatial system. Generally identifying a Type 6 is closely associated with nearby or connected Type 5s and Type 9s, along which Type 8s would often be laterally placed.

It has been suggested that chich mounds (low piles of rubble) might be the foundations of (perishable) buildings (Magnoni et al. 2012). Indeed, the placement in association with building groups of (circular) chich mounds is conspicuously alike the round buildings mapped as architecture. Therefore, in revised iterations, such round buildings have been treated as outlines of buildings, unless their situation within their surroundings seemed to suggest otherwise or their shape seemed illogically irregular for a structure. As a rule of thumb chich mounds with dimensions similar to round buildings and placed detached from albarradas—ancient buildings do not typically straddle albarradas (Magnoni et al. 2012)—are identified as Type 1s.

5 Winchester Materials

In contrast to a lost urban tradition like the Maya, British urban history appears remarkably well-known (see Palliser 2000; Clark 2000; Daunton 2001). A wealth of historical records is typically available to retrieve various narratives behind British urban settlements, often even in considerable detail. However, as many cities are ongoing places continuously being developed over time, historians have struggled to put the intricate structure of material urban space into their research of the past (Bisschops 2012). Methods of reconstructing bygone phases in urban development on the basis of (property) records have been successfully developed since the 1980s (Keene 1985; Keene and Harding 1987). More recent projects are constantly improving the accuracy and detail of these kinds of studies (Lilley et al. 2005, 2007; Lilley 2011a; Lilley Mapping Medieval Chester; Gauthiez and Zeller this volume) by employing new digital technologies. However, Dean (2012b) has recently demonstrated that the established urban morphologically based practice of reconstructing and conjecturing an urban plan sequentially may contain unexpected significant errors, which come to light when combining these mappings with archaeology. Users of historically derived conjectural sequence plans should therefore be critically aware of the inherent and inevitable discrepancies introduced by the methods producing these plans.

The fact that Winchester is still a small but thriving present-day city, confronts the boundary mapping methodology with a different dimension in which to work. Keene's (1985) seminal work on medieval Winchester[7] provides a basic level of urban mapping which can serve as a basis for the detailed resolution boundary mapping requires. Keene (1985) produced three sequences for the Middle Ages: 1300, 1417, and 1550s, on the basis of records of medieval properties or burgage plots. To demonstrate the principle of BLT mapping through time, here only a small section is taken back to 1550s.[8] This is complemented by the current (2011) MM data for Winchester and OS 1872, published during 1871–1872. Needless to say, there are various highly detailed maps from the 1870s onwards, which could give a much greater temporal resolution. This temporal specification could be the subject of future research.

The acquisition of MM is completely digital and can easily be imported into ArcGIS proprietary vector formats. OS 1872 is also acquired digitally, but as a raster image geoTIFF (indicating a basic level of geoprocessing is performed on the image, projecting, locating and scaling it). The Keene plans did not exist in any digital format prior to this research. The original plans appear in sections at a

---

[7] Biddle's 1976 edited volume on early medieval Winchester does not contain mapping material to a similar level of detail, which reflects the increasingly fragmentary nature of the archaeological and historical records required for sequence mapping.

[8] The 1750 Godson survey was also considered for preparing an additional time-slice. However, after appropriate digitisation and georeferencing of this plan from the two copies in the Bodleian Library's collection, it soon transpired that not only the historical survey technology, style of depiction, and imprecise edge matching of the printed sheets, would make a topographical challenge. Also the detailing the detailing of the plan left many building and plot details ambiguous. This rules out the opportunity for a direct translation into realistic and accurate individually reconstructed topographical features. Although the Keene (1985) plans also lack part of the a priori level of detail required for BLT mapping, his two large comprehensive volumes give much detailed background on how to interpret the plans. It is possible that with appropriate historical research, the Godson survey could make a suitable basis for an additional historical section of the city in the future.

1:2500 scale. Although these could be scanned and stitched together, it seemed worthwhile to trace down the originals. These are kept by the Winchester Excavations Committee (curated by Martin Biddle and Katherine Barclay and in the care of the Winchester Research Unit) in the depot of the Winchester City Museum. The originals consist of large sheets of film on which the line drawing of the map was draughted. Rather than sectioned in the small portions of the book, these sheets represented the medieval city in five parts: the walled area, and the north, east, south and west suburbs. The large scale of these original sheets as well as the less fragmented and unannotated nature of them, seemed qualitatively more advantageous to work with than scanning the images in the book. Digitising involved large roller scans of each of the map sheets at high (400-600 dpi) resolution, resulting in huge raster files. Despite the initial advantages of using these large film sheets, the consequential large file sizes caused various issues in the subsequent processing. This involved the cleaning up, enhancing and precise stitching of the five different sheets in Adobe Photoshop. To keep the files manageable a resolution of 400 dpi[9] was deemed sufficiently sharp for the definition of the line drawing at full scale.[10]

The next stage of importing and georeferencing them in ArcGIS was carried out in direct reference to MM, assuming the current mapping standard to be by-and-large the best available. This was preferred over setting up a proper set of control points with dGPS (cf. Lilley 2011a), as error margins between these points could cause unwanted discrepancies between MasterMap and each historical layer. Instead, with the aid of Keene (personal communication) historically persistent points in the environment were identified, documented as spatial data, and photographed on site for future reference. Logically fewer points were available throughout all periods from the 1300s as the built environment developed. These

[9] The originals produced with the help of Geoff Denford, Winchester City Council (Winchester Museum Service), were at 400 and 600 dpi, while additional scans on a larger scanner were made at the University of Portsmouth thanks to Katherine Barclay at 500 dpi. Although the quality of definition on the 500 dpi scans was superior, the lower resolution determined the quality of the final stitched scans, which were visually improved with image processing and filtering in Photoshop thus ensuring readable solid lines, suitable for semi-automated vectorisation (see below).

[10] It should be noted that the Keene plans were prepared in reference to the then current OS city plans of the 1970s, which used planimetric technology closer to present standards (Keene, personal communication). In addition the 1872 OS plan and the 1750 Godson survey were used as points of reference for shaping features in the built environment. The film sheets were all in relatively good condition, but there is no accounting for any errors resulting from 40 years of ageing of the physical material, although flat roller scanning should ensure an accurate reproduction of their current condition without photographic lens distortion. Finally, accuracy is compromised by the stitching process, which is a simplistic visual weighting of the matching errors between the edges of each sheet using graphical tools in Photoshop, which inevitably retains small mismatches. An alternative would have been to vectorise the images and use ArcGIS computational tools to match the edges of the matching vector files.

points served as a basic set of control points,[11] also for the 1872 OS plan. In the georeferencing process the basic control points would achieve enough initial accuracy to carefully select a series of additional points that were clearly related to the historical plan at hand. This improved the relative accuracy, using higher order georectifying warps.

For the 1550s plan, the approach chosen first involved vectorising the raster with a semi-automated process using the ArcScan suite in ArcGIS. Although this also features a fully automated tool, the results of this process left much to be desired. The semi-automated tracing of a two-tone raster image (i.e. classified in two categories) with polylines works somewhat more quickly than an entire manual redraw. The downside is that the thickness of the scanned lines can cause minor ruggedness in the shape of the resulting polylines. For future reference, Keene's (1985) historical conjectures were vectorised in a separate dataset from the features that were deemed certain. Rather than georeferencing, the process of geographically relating the vector data to another dataset is called 'spatial adjustment', but operates on largely similar principles to those in raster georeferencing. Because snapping exactly onto nodes is possible in both the vectorised 1550s data and OS MasterMap, a more accurate placing can be achieved. Any certain identical points can subsequently be selected as 'identity points' where ArcGIS would virtually nail the overlaying sheet to the underlying location. Using rubbersheeting, greater local accuracy on the basis of additional points can be achieved without ever moving these 'identity points'. In this way the vectorised 1550s data could create a more precise initial match to the OS MasterMap data.

5.1 Winchester Base Plans

To demonstrate the applicability of BLT mapping through time and on historical material, a small area around Winchester's former East Gate and bridge was chosen to test the principles. The current MM version[12] represents the large scale

---

[11] In addition, the geographical representation of listed buildings and monument sites was obtained from Winchester City Council (courtesy of Ian Scrivener-Lindley and Tracy Matthews). These polygons and points were prepared on the basis of MasterMap, and so would relate exactly to the source data. Unfortunately, heritage listings serve a policy purpose of protecting and managing the sites. This means their shapes cannot be trusted to convey any historical reality. Furthermore, in Winchester the heritage records, often of a dubious and dated standard, have not been fully integrated across the various systems that have existed over the years and do not include archaeological excavation plans. Only limited cautious use could be made of these records, using online resources such as Heritage Gateway and National Heritage List for England. In practice, where possible, Keene's (1985) words were preferred over the listings, but the records were used to indicate plausible historically persistent features.

[12] MasterMap can be updated up to every six months. This data was downloaded end of October 2011, with the Ordnance Survey Imagery Layer (OS official aerial photography) and Ordnance Survey Address Layer (2) arriving on disc in April 2012.

mapping standard for the UK. Nonetheless, using MM for creating outline features and eventually identifying BLTs is perhaps less straightforward than one would expect. As is the case with all maps MM presents a selection of mapped features. MM should satisfy policy and legal use requirements as well as depicting the physical layout of a place. It omits entrances, and many comprehensive buildings consist of several polygons of which it is unspecified as to how they compose an internally divided whole (as opposed to usual archaeological ground surface mapping). Although the OS Address Layer 2 will give an indication of the location and number of addresses at an approximate location, which helps the interpretation of the physical and social reality, this does not generate the aggregates of polygons which represent each building completely. MM combines a mix of uses in one map, keeping track of a feature's development, extensions and adjustments. It offers a very basic and generalising degree of land use classification, and will often, but not always, indicate the provenance of a feature as either 'natural' or 'man-made'. At the same time most man-made open surfaces are merely described as 'multi surface' or 'general surface', which does not reveal much of what is actually mapped.

This illustrates that even when working on the present, interpretive rules of thumb are necessary to work with the mapped material. Fortunately, we have more resources at our disposal which can shed light on the present social reality of the physically built-up city. MM was used in conjunction with Google Street View, Google Maps, Bing Maps,[13] and Ordnance Survey's Imagery layer (vertical aerial photography). Although this can clarify a lot of what is represented in MM, including revealing minor discrepancies, there are still aspects of the built environment that are largely inaccessible. This mostly concerns the backs of buildings and their gardens, or small alleyways, or legally and functionally restricted areas. For comprehensiveness only a dedicated urban survey might be able to remove most gaps left despite cross-referencing various sources. Again, to acquire an immediate feel for the quality and appropriateness of the data, it is enlightening to perform an initial run on semi-transparent paper.

Creating outlines on the basis of MM required intensive cross-referencing of various sources—photographic sources often being the most intuitive—to choose the lines which truly conveyed outlines that physically exist and are not part of internal design or composite functions in any area. In exceptional instances original MM lines received minor amendments to more precisely convey the actual physical difference on the ground. The greatest ambiguity is undoubtedly associated with separating buildings by internal divisions and likewise, with complex plots and open areas at the back. At the same time, the general assumption held that in inaccessible areas all lines of MM would be physically recognisable on-site, so they could all potentially be used to convey outlines. Inevitably, some of the

[13] Online mapping and imaging resources can be updated without prior notice. The work on Winchester took place between May 2012 and April 2013.

outline features on the base plan were therefore interpreted with potential BLTs in mind. MM itself is topologically integrally developed by the Ordnance Survey. Nonetheless, the tracing of lines in MM is a manual process, so as described, before a topology rule set should be enforced on the outline base plan.

The georeferenced OS 1872 plan likewise introduced its own interpretive difficulties and ambiguities. First OS 1872 was prepared and published over a two year period in which the city was rapidly developing. The seams of the separate sheets of the plan therefore do not always match both due to their separate publication and small differences between the original paper documents. In vectorisation these discrepancies are interpretively weighted to permit mapped features to retain relatively regular shapes. The resolution and definition, as well as the detailed symbology of the OS 1872 plan made it unsuitable for using the semi-automated raster tracing in ArcScan.

Unfortunately, digitised historical OS plans are not delivered with a legend of the symbology and abbreviations used. Although Oliver (1993) mentions the existence of coloured versions of the OS 1872 city plan, these were not available via EDINA's Historical Digimap services. This means the plan is a simple black and white affair, where it is often ambiguous as to what kind of (physical) distinction is made with each single solid line. Coloured plans normally convey differences between built-up areas and open areas as well as to a degree the materials used (Oliver 1993). From comparisons between maps of the same era at the same scale (several can be consulted online at the National Library of Scotland) and an extensive list of abbreviations used in various OS mapping projects over the years (available online at the National Library of Scotland), intensive studying of OS 1872 leads to relatively accurate reading.

OS 1872 was clearly aimed at a comprehensive representation of the physically present features of the city. The general resolution for detailing was 15 cm (Oliver 1993), which offers greater architectural detailing than MM. In addition, functional furnishings of the city were often also mapped. Strangely, contrary to Oliver's (1993) supposition, gates and doorways are not consistently featured on OS 1872, while archways (in walls) do appear. Vectorising towards an outline base plan thus involves selections and interpretations (e.g. excluding the furnishings and some architectural details). As with MM, accuracy cannot be guaranteed for areas around the back of buildings or within larger building complexes which are too compositely mapped to make certain inferences on what each line conveys. Likewise, separately mapped extensions were interpretively incorporated or divided into discrete buildings with internal divisions. Outbuildings are particularly complex as they are numerous in the Victorian city. Rather than each separate rectangle, clusters of outbuildings were given an area outline. As vectorisation relied on manual editing, features of OS 1872 that came tantalisingly close to an extant line in the MM base plan were traced from the latter, which would be deviated from when different shapes and orientations occur.

The base plan of 1550s firstly comprises the spatially adjusted vectorised plot based property plan itself, including Keene's historical conjectures. This dataset offers a much more simplistic foundation, as no unnecessary or confusing detail was included in the historically self-selective reconstructive mapping process. The challenge here is rather the reverse. The limitations of the topographical reconstruction on the basis of the historical records (see Keene 1985; cf. Bisschops 2012) cause many unaffordable gaps inhibiting it to serve as an outline base plan for BLT mapping. Most conspicuously, buildings are not mapped at all, except for those with public and administrative functions. More importantly, as the plans are property based, the physical distinctions of any further subdivisions within any property are not mapped and sometimes simply unknown. This means a rather crude level of conjectural mapping is required to add the missing detail of the built environment to merge into a comprehensive outline base plan.

Keene's (1985, Fig. 155) work conveniently includes a plan for each historical snapshot selected, indicating the built-up and probable built-up frontages along the streets. This is used as a basis for deciding whether a building needs to be conjectured. As for the actual shape of a building there is no pretention that this reflects reality. Lewis et al. (1988) book *Medieval Hall Houses of the Winchester Area* shows three examples of shops surveyed in the city of Winchester, which were between approx. 10 and 15 m in length. These dimensions are taken as a rough maximum for typical buildings, also comparing internally to the more detailed knowledge of smaller separate properties along the High Street area. Without any readily usable direct sources to base morphological considerations on, at this stage it was deemed more important to ensure that the topological distinctions were made than to be concerned about their appearance. Keene's abstracts of compiled historical records on the properties in his gazetteer are used to find clues about any multiple buildings, plots or gardens being part of a single property. Often the evidence for this is scant or even entirely absent (which is also the origin of some of Keene's own conjectures).

As an example, Fig. 6 shows the clear difference between the west and east sides of the northern end of current Chesil Street. The west features large subdivisions on sizeable plots, because no further evidence was available and the suggestion was raised that this area could have hosted a few substantial medieval buildings. The east, however, has been subdivided into smaller built environment features, because there was no special indication of substantial buildings and the one historical building still in existence (The Old Chesil Rectory) was indicated to feature two tenements with a probable communal arched entrance (Keene 1985). The neighbouring plots in that sequence feature frontages (probably built-up, according to Keene) with comparably dividable dimensions (4 or 5 m each). On the opposite corner towards the north, there is an indication that at some point during the late medieval period there could have been six shops occupying this site. In these cases, the open areas behind the buildings are not subdivided as they could well have been shared.[14] Open areas are only subdivided if prompted by Keene's (1985) discussion of the records.

[14] Little is known about the actual (physical) subdivisions of open areas associated with buildings in the medieval period. Archaeologically there could have been fences, paths and

Fig. 6 An example area of the Winchester 1550s base plan data with initial building and plot division conjectures shown in *pink*. The great contrast between the large and small buildings is a reason for concern about its historical accuracy, but is still in keeping with the scarce information available for this area. (Image prepared on the basis of original plans reproduced courtesy of the Winchester Excavations Committee ©.)

Despite the conjecturing efforts, the exact same level of detail cannot be expected on the basis of 1550s as from either MM, OS 1872, or an archaeological plan. It would be a gross over-interpretation to start conjecturing absent outbuildings or morphological details. This means that in any comparative work with more detailed material, the simplified composition of buildings and open areas should be taken into account.

All base plans will need to be checked and corrected on topology before being used for BLT identifications.

5.2 Winchester BLT Identifications

There is no need to repeat the explanation on the general process of BLT identifications as this was described for Chunchucmil before. No fundamental differences occur with a historical European example (see Figs. 7, 8, 9). The rules of thumb applied, however, depend on the particular context and nature of the source material.

(Footnote 14 continued)
hedges, all used to section off small bits of space. In any case, it seems likely the medieval city saw a variety of plot divisions and shared open areas (Dean, personal communication), which is also suggested throughout the discussion of properties in Keene's (1985) gazetteer.

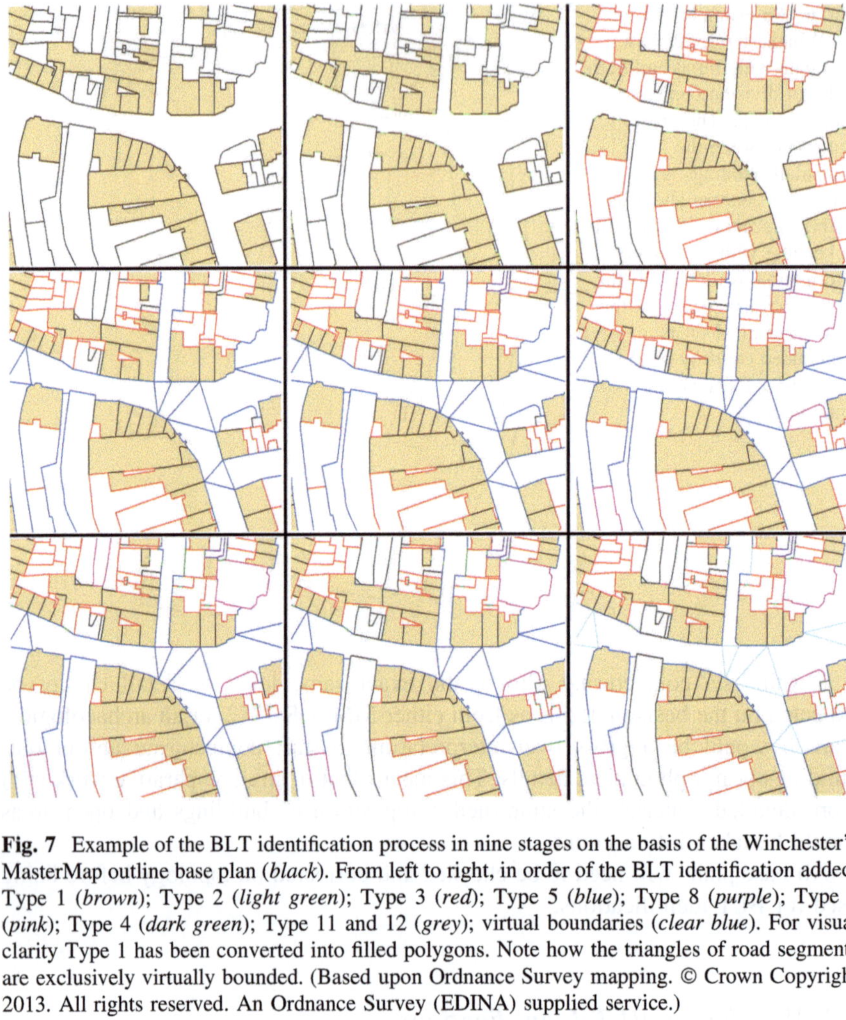

Fig. 7 Example of the BLT identification process in nine stages on the basis of the Winchester's MasterMap outline base plan (*black*). From left to right, in order of the BLT identification added: Type 1 (*brown*); Type 2 (*light green*); Type 3 (*red*); Type 5 (*blue*); Type 8 (*purple*); Type 9 (*pink*); Type 4 (*dark green*); Type 11 and 12 (*grey*); virtual boundaries (*clear blue*). For visual clarity Type 1 has been converted into filled polygons. Note how the triangles of road segments are exclusively virtually bounded. (Based upon Ordnance Survey mapping. © Crown Copyright 2013. All rights reserved. An Ordnance Survey (EDINA) supplied service.)

One aspect particular to working diachronically is to ensure that in the few cases where a boundary outline would remain in exactly the same location through time, any concurring BLT identification on that location should be identical to the more recent phase. This helps to keep the data clean by eliminating confusing uninformed differences. Most instances concern historical building frontages, which also retain the same doorway.

As noted before, the areas around the back of buildings cannot satisfactorily be assessed. This means there is little secure information on back entrances in each time-slice. In general the assumption is then made that back entrances are necessary for structures which have a plot around the back. Unless the shape and

Fig. 8 Example of the BLT identification process in nine stages on the basis of the Winchester's OS 1872 outline base plan (*black*). From left to right, in order of the BLT identification added: Type 1 (*brown*); Type 2 (*light green*); Type 3 (*red*); Type 5 (*blue*); Type 8 (*purple*); Type 9 (*pink*); Type 4 (*dark green*); Type 11 and 12 (*grey*); virtual boundaries (*clear blue*). For visual clarity Type 1 has been converted into filled polygons. (Image prepared on the basis of original scans © Crown Copyright and Landmark Information Group Limited 2013. All rights reserved. 1872.)

mutual orientation of the outlines and further identified BLTs suggests differently, the back entrance is conjectured in rough opposition to the front entrance. Only in complex buildings and contexts or on the basis of actual additional information, would more entrances be identified. It should be noted that because many entrances are conjectured, like in the Chunchucmil case study, their dimension is less relevant than their existence topologically. The width of entrances is at best indicative.

Working on urban built environments more familiar to the researcher means that there is a greater immediate understanding of what could be expected. This applies especially to outbuildings, garages, sheds, follies, etc. (absent in 1550s).

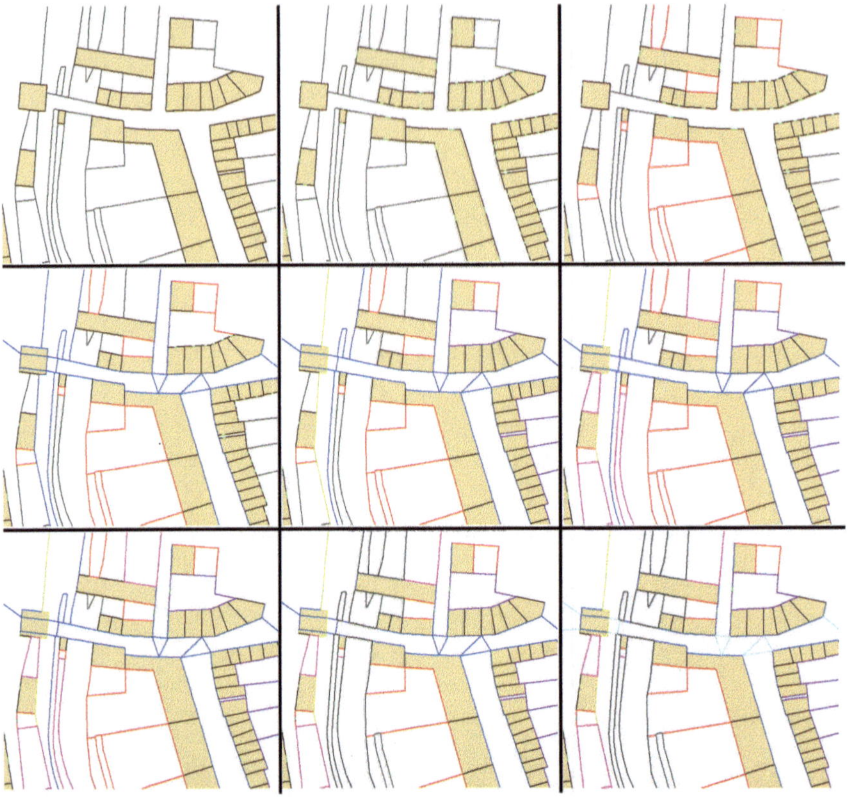

Fig. 9 Example of the BLT identification process in nine stages on the basis of the Winchester's 1550s outline base plan (*black*). From left to right, in order of the BLT identification added: Type 1 (*brown*); Type 2 (*light green*); Type 3 (*red*); Type 5 (*blue*); Type 7, 8 (*yellow, purple*); Type 9 (*pink*); Type 4 (*dark green*); Type 11 and 12 (*grey*); virtual boundaries (*clear blue*). For visual clarity Type 1 has been converted into filled polygons. Note the effects of the city wall (Type 7) and the medieval encroachments on the river along the bridge. (Image prepared on the basis of original plans reproduced courtesy of the Winchester Excavations Committee ©.)

In Chunchucmil all architecturally mapped structures and architecture ended up with Type 1 identifications, while at the same time many actual outbuildings could have perished. This may have included examples of what in western and globalised cities would be considered (functional) outbuildings. However, in Chunchucmil and indeed the Maya culture area in general, buildings tend to occur in groups on shared open areas. These open areas would usually be identified as a Type 3 or a Type 8, depending on a one-to-one or one-to-many building(s) relationship. In western and global cities outbuildings are typically understood as auxiliary to and under the exclusive influence of the main structure they are in constellation with (a single socio-spatial system). It therefore seems detrimental to go against that intrinsic understanding and consequently identify most gardens as a socio-spatially distinct Type 8, while there is still only one main Type 1 which determines the association.

This is why outbuildings in Winchester (MM and OS 1872) are treated as sequential occurrences of Type 3s or 8s even if materially they could be closable like separately inhabited or occupied structures. In Chunchucmil the groups of Type 1s are typically larger than two and extensive research suggests that they are inhabited by multiple nuclear families (Johnston and Gonlin 1998; Magnoni 2007), meaning that with greater confidence their associated open areas justifiably end up a Type 8. The advantage of this practice is that visually and computationally the BLT data is still easily simplified to generate aggregate outlines of a Type 3s or 8s cluster, which could stand on the same level as badly preserved archaeology or self-selective historical reconstructions. Since it might be more difficult to repeat the entire interpretation process to add detail later, maintaining auxiliary buildings is preferable to omitting them initially. Additional information from other sources could in any future research be stored in the attributes of the BLTs as well.

Boundaries of unoccupiability (Type 11 or 12) are primarily based on MasterMap's proprietary 'natural' or 'man-made' designations. On OS 1872 they are based on the symbology, and on 1550s they are limited to bodies of water only.

Due to the style of OS 1872, the nature of outlines often does not reveal itself clearly until the BLT identifications disentangle all the lines. This means that similar to the iterative revisions of conjectures on Chunchucmil's plan, the OS 1872 base plan is more invasively revised during the BLT identification process than is the case for either MM or 1550s. At the same time the additional architectural detail and pathways in parks and gardens available on OS 1872 only, give an extra certainty for conjecturing any entrances.

Garden plots situated like housing plots without a directly associated building on site are quite particular to the medieval period in Winchester. Though justifiably identified as opening boundaries, Type 9s, they are something of an oddity. Due to their open character these gardens are logically a Type 9, but the known function is quite distinct from parks or other open areas. This difference is similar to the distinction of an agricultural field and a park, which resonates well considering garden plots could have been used to grow crops rather than serving a more modern leisure function. Besides built-up frontages, Keene (1985: Fig. 155) also identifies likely 'open ground', which seems to indicate land without any particular identified use, which roughly follows the general plot pattern. With the current BLT ontology it is impossible to differentiate between these and so-called garden plots. Despite differentiation of what is typically regarded as open space in urban studies (Stanley et al. 2012; Smith 2008), functionally Type 9s will have a protean referral record. Although spatially often unidentifiable, from a social perspective the ambiguity due to the lack of a predominant socio-spatial occupation, could justifiably render any unused space (e.g. fallow) a Type 10.

6 Interpretative and Descriptive Merits

Now we have seen the application of BLT mapping in an archaeological and alternative urban tradition as well as in a historical example of an ongoing urban tradition, a brief discussion of the merit of this intensive mapping method is in order. This will be done by focusing on the identification of one BLT: Type 8 (mutual boundaries).

Type 8's definition pertains particularly to the nature of an area that is formed and occupied by an aggregate socio-spatial system composed of a specific selection of materially secluded socio-spatial systems (occupying physically segregating structures). As noted before, Chunchucmil's urban environment, and indeed Classic lowland Maya urbanism, is characterised by house or building groups in a variety of constellations. It is therefore hardly surprising that Type 8s end up as a part of a large number of well-dispersed sizeable areas. However, BLT mapping not only confirms on a spatial and physical level what can generally be asserted by looking at plaza and platform groups, it also makes explicit how they are composed and how they are connected and situated on a social and spatial level within the urban environment as a whole. It specifies that only very few open areas pertain singularly to a structure (Type 3) within Type 8s. Adjacent Type 8s furthermore create larger clusters of groups together with their shared spaces, around which, by and large, circulation spaces guide further open interaction opportunities. It also demonstrates that plazas and platforms, distinct from Type 8s as identified in many plaza groups, should in the majority of contexts not be confused with current ideas about urban squares, which often form locally central parts of circulation. Instead, that role often appears to be played by opening boundaries (Type 9s), while square like areas (Type 6s) are a much more specific occurrence.

Type 8s in Winchester are of interest, initially because comparing with Chunchucmil, present-day Winchester is largely devoid of Type 8s in favour of a multitude of Type 3s. When they do occur, they are often set back from the more open circulation systems. Their boundaries therefore only represent the first point of interaction with the secluded socio-spatial systems (Type 1s) grouped by the Type 8 in a restricted way. This latter observation holds true for both OS 1872 and 1550s (Figs. 8 and 9). Both these historical situations show a mix of a Type 8 and Type 3 pattern, where in OS 1872 fewer Type 1s participate in Type 8 constellations. Type 8s also appear marginally less frequent, yet possibly more evenly distributed than on 1550s.[15] OS 1872 is therefore suggestive of a transitional stage in which associated open spaces become increasingly secluded.

This preliminary observation of a possible urban historical process in Winchester's built environment, uncovered by the BLT identification process, benefits from an analogy with historical research to test the reliance on exclusively empirically mapped information. Daunton's (1983) spatially sensitive quotidian

[15] As noted before, this could partially be due to the way the historically reconstructed map and lack of documentary detail was treated in the outline base plan.

study of the form and formation of the Victorian city, confirms that the careful reading of OS 1872 can be trusted. The British Victorian city saw a marked shift from shared open areas to increasingly individual open areas, even to the point where policy measures were put in place for the construction of urban residences including such open areas. Although Daunton's discussion focuses on an area of Newcastle and although the changes emerged more quickly in various guises in large urban settlements, the move towards more privately allocated open space instead of accessible open meeting spaces seeped into the provinces over time (Daunton 1983). It is therefore very likely that 1870s Winchester was a Victorian city in transition. The socially interpretive and descriptive use (cf. Griffiths 2013) of the BLT identifications comes into its own here. The virtual boundaries, for example, give the researcher the interpretive flexibility to mark the difference of an incomplete separation between individual buildings' open areas. They offer a stage in between Type 8s (mutual boundaries) and fully fledged Type 3s (associative boundaries).

For all of these socially positioned and spatially situated occurrences of Type 8s, or indeed any BLT identification, applies that their description is only completed by all other BLT identifications of which they are part. They are incrementally contextualised within a completely marked-up dataset of a built environment. Despite bringing to bear a novel essentialist ontology onto the built environment, it could be argued that, due to superior understanding of western and globalised society and its urban forms, this method adds little to the initial socio-cultural understanding of the structure of a particular urban space. The advantages offered by BLT mapping should rather be sought in the opportunities it creates for the resurging research interest in comparative urbanism (Smith 2012; Smith and Peregrine 2012; Stanley et al. 2012; Taylor 2012; York et al. 2011; Ward 2010; Fletcher 2009; Yoffee 2009; McCafferty and Peuramaki-Brown 2007; Nijman 2007; Briggs 2004; Robinson 2004).

The enigmatic and irregular nature of the Maya urban tradition has caused havoc to their study as cities (Sanders and Webster 1988; Chase et al. 1990; Graham 1999; Smith 1989, 2003; McCafferty and Peuramaki-Brown 2007). Although the idea of low-density urbanism has existed for decades, it is only recently that archaeological examples like the Maya are receiving more structural attention directed at the analysis of the pattern of urbanism they represent (Fletcher 2009; Isendahl and Smith 2013; Arnauld et al. 2012). As demonstrated here, the boundary mapping method is able to contribute to the opportunities for analysis referring to urban form or the configuration of the built environment in particular. In the Maya area it offers a mapping which could further formalise, specify and spatially contextualise work on neighbourhood clusters (Ek 2006; Hare and Masson 2012; Lemonnier 2012) or political modalities (Bazy 2011). Furthermore, its topological data structure (see below) intrinsically provides the basis for many additional lines of questioning spatial patterns and connectivity (cf. Morton et al. 2012b, this volume, applying space syntax), or further specifying and investigating possible alignments across Maya settlement layouts (cf. Bevan et al. 2013). In addition, the outline and BLT specifications, open further directions for discussing Chunchucmil in relation to Becker's (2001) settlement layout models (cf. Magnoni et al. 2012) and attributed with additional material and architectural information, this could be connected to considerations of labour investment (Folan et al. 2009; Guderjan 2007).

Thus BLT mapping allows initial explorations of alternative urban traditions where extant methods have difficulty engaging with the structure of their built environments as they were developed for globalised cities. The boundary method is devised for it to work equally across all built environments. Despite this flexibility, it fulfils the requirement of a more rigorous and formal method on the basis of a directly empirically translated theory (Smith 2003, 2011a; Smith and Peregrine 2012, Joyce 2009), which sits next to perhaps overly formal and socio-culturally embedded or prescriptive current methods to study urban form. A western medieval city like Winchester, relying heavily on a street network and a Maya city composed of intricately linked series of variably characterised and embedded open space are descriptively and formally mapped on exactly the same basis. This allows for a meticulous exploration and description of the socio-spatial affordance of inhabitation in physically developing configurations of subdivisions, contextualising and situating each specific location on the same grounds.

7 Diachronic and Spatial Data Structure

Bisschops (2012) rightly notes that digital maps of contemporary urban situations are not suitable for plotting and attributing medieval historical data onto them. These modern maps contain much irrelevant information and fail to represent the spatial situation for the historical moment of interest. Therefore these practices offer little opportunity to question the relationship between society and space in the past, let alone elucidate socio-spatial processes through time (cf. Griffiths 2013). Relating the built environment of the past rigorously to the present, however, would enable a study of the social materialising processes of the mutual constitution of the city and its inhabitants.

For comprehensive visual inspection alone, the images throughout this paper have demonstrated the need for a better way to visualise the mapped BLT information holistically. Naturally, where the data available for a case study allows, the resultant data structure emerges diachronically and thus could be compared and analysed across time. Figure 10 shows the outline base plans for the same area in Winchester as used before for the three time-slices (1550s, green; OS 1872, blue; MM, red). Where most diachronic images in historical GIS are raster based and thus require semi-transparent layers on which point data of events can be distributed (e.g. Amherst College 2009; Amherst College 2010), here the vector topology of the data simultaneously provides both a spatially clearer and perceptively more complex image. It is easier to see specific relations in small areas than it is to get an overall impression of the processes going on. If this is true for

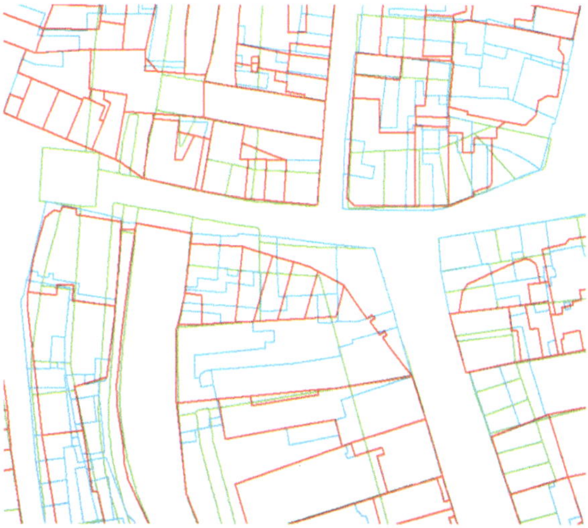

Fig. 10 All outline base plans are overlaid on top of each other in the same Winchester area: MM (*red*); OS 1872 (*blue*); and 1550s (*green*). Note that the major changes from 1550 to 1872 are formed by the removal of the city wall, widening of the bridge, and the intensification of built-up space, e.g. along the river. In addition, it can be seen how the major changes from 1872 to the present concern infrastructural adjustments to street lines, some major new buildings and a cleaning up of the mishmash of development along the west bank of the river. At the same time some of the plot boundaries along Chesil Street are amongst the most persistent features. (Partly based upon Ordnance Survey mapping. © Crown Copyright 2013. All rights reserved. An Ordnance Survey (EDINA) supplied service; partly based upon original plans reproduced courtesy of the Winchester Excavations Committee ©; partly based upon original scans © Crown Copyright and Landmark Information Group Limited 2013. All rights reserved. 1872.)

the relative simplicity of three time-slices, it would most definitely require other measures when thirteen BLT layers per time-slice are added.

Importantly, however, the BLTs convey what occurred in the composition of the city's built environment itself and how the development of this has manifested itself as is relevant to and constitutive of its inhabiting urban society. Conceptually and experientially the structural composition of the built environment does not adhere to the graphic area pixels of a raster image, but naturally relates to topological relations along lines forming composite subdividing polygons (areas). Deriving specific spatially contextualised insights on urban configuration formation requires a vector GIS data format. Yet, working with BLTs through time effectively necessitates advanced spatial analysis and associated visualisations to obtain significant information coherently.

By conducting the mapping processes in a GIS, but on a conceptual basis, an intricate spatial data structure emerges. This data structure would not normally be at the disposal of historical or social scientists studying material characteristics of urban space together with the relationships between society and space at large. It offers the flexibility to incorporate various degrees of socio-cultural, functional, or

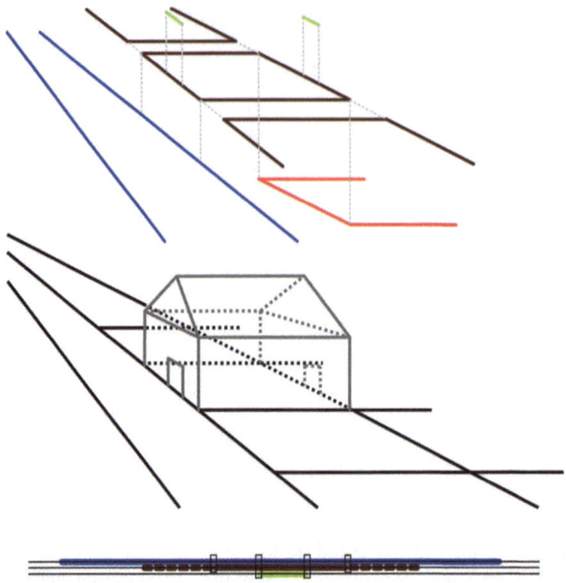

Fig. 11 A schematic representation of the GIS data structure resulting from conducting BLT mapping. The 3D terraced house (*grey*) on a street represents a simple empirical reality in a usual globalised urban built environment. The base plan (*black*) results from the physical subdivisions recognised on ground level. In colours above all BLT identifications that would operate on the rectangle of the house in the depicted situation are shown in layers: Type 1 (*brown*); Type 2 (*green*); Type 3 (*red*); Type 5 (*blue*). Note how only the identifications of the Type 1 of the house's outline and its entrances are completely drawn. Below, the boundary outline of the front of the house is presented as overlaid GIS layers (in thin black lines). The rectangles represent all the nodes operating at the front of the house. Note how three complete socio-spatial descriptions of boundaries emerge: the doorway itself, left of the doorway, and right of the doorway. All are topologically distinct and unique boundary segments

time–space historical specification as attributes in the GIS, moving into ever more casuistic applications. This can be achieved on the basis of three units: either per BLT identification, per completely described subdivision, or per specific BLT combination at each unique topological segment (see Fig. 11). Such attribute specifications would never detract from the intrinsic socio-spatial significance conveyed by each spatial data element.

The redescribed built environment plan falls apart in the combinations of colocated BLT identifications along each unique segment of boundary outline. These smallest elements pertain to an essential understanding of the social significance of the presence of that distinction as a site of difference (cf. Abbott 1995). The occurrence of each combination of BLTs gives rise to an additional emergent categorisation characterising a chosen urban built environment. An initial downside is that rather than representing readily perceived socio-cultural or functional spatial categories, these pieces of line complicate the image of the built environment we have, despite allowing precise socio-spatial assessment. Tools for the spatial analysis of these new layers of significance within our data are needed to aid us in exploring large and complex built environment configurations and extract any patterns emerging from the intrinsic characteristics of the particular built form. The analysis of the rhythms and patterns in the composition and interconnectivity created by the smallest meaningful elements of topological segments could reveal aggregate levels of coherence inherently present in a city.

8 Future Potential of BLT Mapping

Despite the need for computational spatial analytical tools to help navigate and order the complexity of the resulting datasets, BLT mapping provides a means to not merely assess the changes and development of urban form *per se*, but the specific socio-spatial characteristics of developing urban forms. That is, the boundary outline itself could change, but even when the outline stays the same, the BLT identifications along the line could change. This offers a key to a more detailed understanding of urban development and to a recognition of socially significant patterns within those processes in each particular case.

An additional advantage of the boundary mapping method is that the resultant spatial data does not abstract or simplify the representation of urban built environments into something that cannot be combined with other extant methodologies. For example, its outline base plans could be adopted as the basis for a space syntactic study, while the intricate data structure could be invested with urban morphological attributes to add layers of information to the research without losing its fundamental socio-spatial bearing. This is important as it has been noted recently that integration of urban morphology and space syntax, or space syntax and GIS, are quite complicated to achieve despite commonalities (Kropf 2009; Pinho and Oliviera 2009; Jones et al. 2009; Griffiths et al. 2010). Although there are undoubtedly various practical complications, the structure of the spatial data generated in boundary mapping seems conducive to integration, because most elementary characteristics of urban form are maintained rather than transformed. Parallel ontologies or typologies of urban tissue and architectural textures (cf. Kropf 1996, 2009, 2011), a functional and scalar typology of urban open space (cf. Stanley et al. 2012), or meaning carrying elements in buildings and planning (cf. Fisher 2009; Smith 2007) could be incorporated, by attributing either topological boundary segments, completely described subdivisions, or individual BLTs.

Flexibility of the method is also found in the source materials used. This paper has demonstrated that in general archaeological survey plans, modern, historical, and historically reconstructed plans can be successfully subjected to BLT mapping. Some of the decisions and assumptions made in this process differ according to that source material, which does not obstruct the method to progress to create fundamentally equal data across cases. Nonetheless, it also demonstrates that the archaeological focus on observing and documenting physically present information makes a better natural fit to this method's requirements and provides important correctives on historical conduct (see Dean 2012b). And within archaeological conduct several alternative mapping practices exist. Not only excavation plans (in other Winchester areas considerable excavation data is available [Keene, personal communication; Scobie et al. 1991)], but also the data produced by various remote sensing techniques reach levels of definition which produce urban plans of sufficiently good quality (see Benech 2007; Maschek et al. 2012; Paliou et al. 2012).

As an analytical mapping practice within a GIS environment acting as research tool, BLT mapping perhaps does not aim to push forward the theoretical basis of GIS in particular (cf. Haçiguzeller 2012; Gillings 2012; Kwan and Schwanen 2009; Leszczynski 2009), but rather forces GIS to work in a qualitative and theoretically informed way (cf. Hu 2011). It is founded on a framework of human and social understanding structurally coupled with the affording nature of material presence (cf. Gillings 2012; see Vis 2013a). It is historically sensitive in both theory and practice. Requiring first the empirical materiality of space through time, the main gist of the spatial turn in historical sciences—reading spaces as produced in a Lefebvrian irreducible and imaginable way (cf. Griffiths 2013; Arnade et al. 2002)—is incorporated in its descriptive and analytical potential. Mapping boundary line types is inescapably a method for the qualitative social scientific use of GIS.

Acknowledgments This research was funded by a University of Leeds Research Scholarship. This work would not have been possible without the kind and patient assistance, dedication and astute insights of the respective map makers and data holders: Scott Hutson, Derek Keene, Martin Biddle, Katherine Barclay, Geoff Denford, Ian Scrivener-Lindley, and Tracy Matthews and the projects or organisations behind them: the Pakbeh Regional Economy Program, the Winchester Excavations Committee (in particular the Winchester Research Unit), Winchester City Council (in particular the Winchester Museum Service). Thanks also to Mark Gillings for helping me sort out digital file conversions. Andrew Evans, Penelope Goodman and Alex Schafran offered useful comments on an earlier version of this chapter. Thanks to Jonathan Sela for helping out with the final details. For any data requests, contact the author.

References

Abbott A (1995) Things of boundaries. Soc Res 62(4):857–882

Amherst College (2009) Tokyo Cityscape. http://ats.amherst.edu/tokyodemo/. Accessed 6 May, 2013

Amherst College (2010) Paris Cityscape. http://ats.amherst.edu/parisdemo/. Accessed 6 May, 2013

Arnade PJ, Howell MC, Simons W (2002) Fertile spaces: the productivity of urban space in northern Europe. J Interdisc Hist 32(4):515–548

Arnauld MC, Manzanilla LR, Smith ME (eds) (2012) The neighborhood as a social and spatial unit in mesoamerican cities. The University of Arizona Press, Tucson

Bafna S (2003) Space syntax: a brief introduction to its logic and techniques. Environ Behav 35(1):17–29

Batty M, Rana S (2004) The automatic definition and generation of axial lines and axial maps. Environ Plan B 31:615–640

Bazy D (2011) Las Modalidades y Dinámicas de las Relaciones entre Facciones Políticas en Piedras Negras, Petén, Guatemala: El dualism politico. Traces 59:59–73

Becker JM (2001) Houselots at Tikal Guatemala: It's what's out back that counts. In: Ruiz AC, Ponce de Leon MJI and Carmen Martinez Martinez M del (eds) Reconstruyendo la Ciudad Maya: El urbanismo en las sociedades antiguas, Publicaciones de la SEEM, no. 6), Sociedad Española de Estudios Mayas, Madrid, pp 427–460

Becker JM (2004) Maya heterarchy as inferred from classic-period plaza plans. Anc Mesoam 15:127–138

Beisaw AM, Gibb JG (2013) Mapping town formation: precision, accuracy, and memory. Presented at: Society for Historical Archaeology annual meeting, 2013, Leicester, UK

Benech C (2007) New approach to the study of city planning and domestic dwellings in the ancient near east. Archaeol Prospection 14:87–103

Bevan A, Jobbová E, Helmke C, Awe J (2013) Directional layouts in central lowland maya settlement. J Archaeol Sci 40(5):2373–2383

Biddle M (ed) (1976) Winchester in the early middle ages: an edition and discussion of the Winton domesday, Winchester Studies 1. Oxford University Press, Oxford

Bisschops T (2012) It is all about location: GIS, property records and the role of space in shaping late medieval urban life. The case of Antwerp around 1400. Post-Class Archaeol 2:83–106

de Briggs XS (2004) Civilization in color: the multicultural city in three millennia. City & Commun 3(4):311–342

Chase DZ, Chase AF, Haviland WA (1990) The classic maya city: reconsidering the 'Mesoamerican urban tradition'. Am Anthropol, New Ser 92(2):499–506

Chiaradia A, Schwander C, Gil J, Friedrich E (2008) Mapping the intangible value of urban layout (i-VALUL): Developing a tool kit for the socio-economic valuation of urban area, for designers and decision makers. 9th international conference on design & decision support systems in architecture and urban planning, Eindhoven, Netherlands. http://www.academia.edu/454889/Mapping_the_intangible_value_of_urban_layout_i-VALUL_Developing_a_tool_kit_for_the_socio-economic_valuation_of_urban_areas_for_designers_and_decision_makers. Accessed 6 May 2013

Clark P (ed) (2000) The Cambridge urban history of Britain, vol II, Cambridge University Press, Cambridge, pp 1540–1840

Conzen MRG (1960) Alnwick, Northumberland: a study in town-plan analysis. Trans Paper Inst Br Geogr 27:iii,ix–xi,1,3–122

Conzen MRG (1968) The use of town plans in the study of urban history. In: Dyos HJ (ed) The study of urban history. Edward Arnold, London, pp 113–130

Conzen MRG (1981) The plan analysis of an English City centre. In: Whitehand JWR (ed) The urban landscape: historical development and management. Academic Press, London, pp 25–53

Conzen MRG (2004) Thinking about urban form: papers on urban morphology 1932–1998. Verlag Peter Lang, New York

Craane M (2009) The medieval urban 'movement economy': using space syntax in the study of medieval towns as exemplified by the town of 's-Hertogenbosch, the Netherlands. In: Koch D, Marcus L, Steen J (eds) Proceedings of the 7th international space syntax symposium. KTH, Stockholm, 019

Dahlin B (2000) The Barricade and Abandonment of Chunchucmil: implications for northern Maya warfare. Lat Am Antiquity 11(3):283–298

Daunton MJ (1983) Public place and private space: the Victorian city and the working-class household. In: Fraser D, Sutcliffe A (eds) The Pursuit of Urban History. Edward Arnold, London, pp 212–233

Daunton M (ed) (2001) The Cambridge urban history of Britain, vol III, Cambridge University Press, Cambridge, pp 1840–1950

Dean G (2012a) GIS, archaeology and neighbourhood assemblages in Medieval York. Post-Class Archaeol 2:7–29

Dean, G. 2012b, Urban neighbourhoods: social and spatial development in York c.600–1600, unpublished PhD dissertation, University of York

Demarest A (1997) The Vanderbilt Petexbatun regional archaeological procect 1989–1994: overview, history, and major results of a multidisciplinary study of the Classic Maya Collapse. Anc Mesoam 8:209–227
Ek JD (2006) Classic maya urbanism: a GIS approach to site planning, demography, and integration, in: 71st annual meeting of the society for American archaeology, San Juan, Puerto Rico
Faulkner M (ed) Mapping medieval chester project. www.medievalchester.ac.uk. Accessed 6 May 2013
Fisher KD (2009) Placing social interaction: an integrative approach to analyzing past built environments. J Anthropol Archaeol 28:439–457
Fletcher R (2004) Materiality, space, time and outcome. In: Bintliff JL (ed) A companion to archaeology. Blackwell Publishing, Oxford, pp 110–140
Fletcher R (2009) Low-density, agrarian-based urbanism: a comparative view. Insights 2(4):1–19
Folan WJ, Hernandez AA, Kintz ER, Fletcher LA, Gonzalez Heredia R, Hau JM, Caamal Canche N (2009) Mexico: a recent analysis of the social, economic and political organization of a major Maya urban center. Anc Mesoam 20:59–70
Frank Z (2013) Layers, flows, intersections: historical GIS for 19th-century Rio de Janeiro. http://vimeo.com/60104031. Accessed 6 May 2013
Franz G, Wiener JM (2008) From space syntax to space semantics: a behaviourally and perceptually oriented methodology for the efficient description of the geometry and topology of environments. Environ Plan B 35:574–592
Galinié H, Rodier X, Saligny L (2004) Entités Fonctionnelles, Entités Spatiales et Dynamique Urbaine dans la Longue Durée. Histoire & Mesure 19(3/4):223–242
Gauthiez B, Zeller O (this volume) Lyons, the spatial analysis of a city in the 17th and 18th centuries: locating and crossing data in a GIS built from written sources
Gibson J (1979) The ecological approach to visual perception. Houghton Mifflin, Boston
Gillings M (2012) Landscape Phenomenology, GIS and the Role of Affordance. J Archaeol Method Theory 19(4):601–611
Graham E (1999) Stone Cities, Green Cities. Archeol Pap Am Anthropol Assoc 9(1):185–194
Gregory IN, Ell PS (2007) Historical GIS: technologies, methodologies and scholarship. Cambridge University Press, Cambridge
Griffiths S (2005) Historical space and the practice of 'spatial history': the spatio-functional transformation of Sheffield 1770–1850. In: van Nes A (ed) 5th international space syntax symposium proceedings. Techne Press, Amsterdam, pp 655–668
Griffiths S (2011) Temporality in Hillier and Hanson's theory of spatial description: some implications of historical research for space syntax. J Sp Syn 2(1):73–96
Griffiths S (2013) GIS and research into historical "spaces of practice": overcoming the epistemological barriers. In: von Lünen A, Travis C (eds) History and GIS: epistemologies, considerations and reflections. Springer, Dordrecht, pp 153–171
Griffiths S, Quick T (2005) How the individual, society and space become structurally coupled over time. In: van Nes A (ed) 5th international space syntax symposium proceedings. Techne Press, Amsterdam, pp 447–458
Griffiths S, Jones C, Vaughan L, Hacklay M (2010) The persistence of suburban centres in greater London: combining conzenian and space syntax approaches. Urb Morphol 14(2):85–99
Guderjan TH (2007) The nature of an ancient maya city: resources, interaction, and power at Blue Creek. The University of Alabama Press, Tuscaloosa
Haçiguzeller P (2012) GIS, critique, representation and beyond. J Soc Archaeol 12(2):245–263
Hare TS, Masson MA (2012) Intermediate-scale patterns in the urban environment of postclassic Mayapan. In: Arnauld MC, Manzanilla LR, Smith ME (eds) The neighborhood as a social and spatial unit in Mesoamerican cities. The University of Arizona Press, Tucson, pp 229–260
Heritage gateway. http://www.heritagegateway.org.uk/Gateway/. Accessed 8 May 2013
Hillier B (2007) Space is the machine. Space Syntax, London
Hillier B, Hanson J (1984) The social logic of space. Cambridge University Press, Cambridge

Hu D (2011) Advancing Theory? Landscape archaeology and geographical information systems. Papers from the Institute of Archaeology, University College London 21:80–90
Hutson SR, Stanton TW, Magnoni A, Terry R, Craner J (2007) Beyond the buildings: formation processes of ancient Maya houselots and methods for the study of non-architectural space. J Anthropol Archaeol 26:442–473
Hutson SR, Hixson DR, Magnoni A, Mazeau D, Dahlin B (2008) Site and community at Chunchucmil and ancient Maya urban centres. J Field Archaeol 33(1):19–40
Hutson SR, Magnoni A, Dahlin B (in preparation) Maya market economy: multidisciplinary research at ancient Chunchucmil
Isendahl C, Smith ME (2013) Sustainable agrarian urbanism: the low-density cities of the Mayas and Aztecs. Cities 31:132–143
Jensen JT, Keyes G (2003) Mapping urban history: GIS and the analysis of the urban space of nineteenth-century Aarhus. International association for history and computing XVth conference in Tromsø. http://www.rhd.uit.no/ahc/paper/jtj_gk_mapping_urban_history.pdf. Accessed 6 May 2013
Johnston KJ, Gonlin N (1998) What do houses mean? approaches to the analysis of classic maya commoner residences. In: Houston SD (ed) Function and meaning in classic Maya architecture. Dumbarton Oaks, Washington, pp 141–185
Jones C, Griffiths S, Mordechay H, Vaughan L (2009) A multi-disciplinary perspective on the built environment: space syntax and cartography—the communication challenge. In: Koch D, Marcus L, Steen J (eds) Proceedings of the 7th international space syntax symposium, KTH, Stockholm, 048
Jones R (2004) What time human geography? Prog Hum Geogr 28(3):287–304
Joyce AA (2009) Theorizing urbanism in ancient Mesoamerica. Anc Mesoam 20:189–196
Keene D (1985) Survey of medieval winchester, winchester studies 2, vol vol I and II. Oxford University Press, Oxford
Keene D, Harding V (1987) Historical gazetteer of London before the great fire: cheapside; parishes of all hallows honey lane, St Martin Pomary, St Mary le Bow, St Mary Colechurch and St Pancras Soper Lane. http://www.british-history.ac.uk/source.aspx?pubid=8. Accessed 6 May 2013
Kropf KS (1996) Urban tissue and the character of towns. Urb Des Int 1(3):247–263
Kropf KS (2009) Aspects of Urban Form. Urb Morphol 13(2):105–120
Kropf KS (2011) Morphological investigations: cutting into the substance of urban form. Built Environ 37(4):393–405
Kwan M-P, Schwanen T (2009) Guest editorial: critical quantitative geographies. Environ Plann A 41:261–264
Lefebvre B (2009) How to describe and show dynamics of urban fabric: cartography and chronometry? In: Proceedings of the 37th computer application and quantitative methods in archaeology conference, Williamsburg
Lefebvre B (2012) The study of urban fabric dynamics in long time spans: modelling, analysis and representation of spatio-temporal transformations. Post-Class Archaeol 2:65–82
Lefebvre B, Rodier X, Saligny L (2008) Understanding urban fabric with the OH_FET model based on social use, space and time. Archeologia e Calcolatori 19:195–214
Lemonnier E (2012) Neighborhoods in classic lowland Maya societies: their identification and definition from the La Joyanca case study (northwestern Petén, Guatemala). In: Arnauld MC, Manzanilla LR, Smith ME (eds) The neighborhood as a social and spatial unit in Mesoamerican cities. The University of Arizona Press, Tucson, pp 181–201
Leszczynski A (2009) Quantitative limits to qualitative engagements: GIS, its critics, and the philosophical divide. Prof Geogr 63(3):350–365
Lewis E, Roberts E, Roberts K (1988) Medieval hall houses of the Winchester area. Winchester City Museum, Winchester
Lilley KD, Digital mappings. In: Faulkner M (ed) Mapping medieval chester project. www.medievalchester.ac.uk. Accessed 6 May 2013

Lilley KD (2000) Mapping the Medieval city: plan analysis and urban history. Urb Hist 27(1):5–30
Lilley KD (2011a) Urban mappings: visualizing late medieval chester in cartographic and textual form. In: Clarke C (ed) Mapping the Medieval city. University of Wales Press, Cardiff, pp 19–41
Lilley KD (2011b) Geography's medieval history: a neglected enterprise? Dialogues Hum Geogr 1(2):147–162
Lilley KD (2012) Mapping Truth? spatial technologies and the Medieval city: a critical cartography. Post-Class Archaeol 2:201–224
Lilley KD, Lloyd CD, Trick S (2005) Mapping medieval townscapes: a digital atlas of the new towns of Edward i. http://archaeologydataservice.ac.uk/archives/view/atlas_ahrb_2005/. Accessed 6 May 2013
Lilley KD, Lloyd CD, Trick S (2007) Designs and designers of medieval 'new towns' in wales. Antiquity 81:279–293
Locating London's past, version 1.0. www.locatinglondon.org. Accessed 6 May 2013
Lynch K (1981) Good urban form. MIT Press, Cambridge
MacEachren AM (2004) How maps work: representation, visualization and design. The Guilford Press, New York
Magnoni A (2007) Population estimates at the ancient Maya city of Chunchucmil, Yucatán, Mexico. In: Clark JT, Hagemeister EM (eds) Proceedings of 34th computer application and quantitative methods in archaeology conference, Fargo, pp 160–167
Magnoni A, Hutson S, Dahlin B (2012) Living in the city: settlement patterns and the urban experience at classic period Chunchucmil. Yucatan, M, Anc Mesoam 23(2):313–343
Manzanilla L, Barba L (1990) The study of activities in classic households: two case studies from Coba and Teotihuacan. Anc Mesoam 1:41–49
Maschek D, Schneyder M, Tschannerl M (2012) The Civilian town of Carnuntum (Lower Austria) in time and space: a multi-layered approach towards the reconstruction of urban transformation. In: Börner W, Uhlirz S, Dollhofer S (eds) 16th International conference on "cultural heritage and new technologies" Vienna, 2011, Museen der Stadt Wien, Stadtarchäologie, Vienna, pp 348–361
McCafferty GG, Peuramaki-Brown M (2007) Ancient cities of Mesoamerica. West Hum Rev 61(3):100–111
Monmonier M (1996) How to lie with maps, 2nd edn. University of Chicago Press, Chicago
Morton SG, Peuramaki-Brown MM, Dawson PC, Seibert JD (2012a) Civic and household community relationships at Teotihuacan, Mexico: a space syntax approach. Camb Archaeol J 22(03):387–400
Morton SG, Peuramaki-Brown MM, Dawson PC, Seibert JD (2012b) The dynamic maya city: methods for modelling pedestrian movement in ancient urban centres. Presented at: 17th European Maya conference, University of Helsinki
Morton SG, Peuramaki-Brown MM, Dawson PC, Seibert JD (this volume) Peopling the past: interpreting models for pedestrian movement in ancient civic-ceremonial centres
Moudon AV (1997) Urban morphology as an emerging interdisciplinary field. Urb Morphol 1:3–10
Narvaez L, Penn A, Griffiths S (2012) Creating urban place: re-thinking the value of residential and commercial use in urban street networks. Sp Flows: Int J Urb ExtraUrb Stud 2(3):149–168
National Heritage List for England. http://www.english-heritage.org.uk/professional/protection/process/national-heritage-list-for-england/. Accessed 8 May 2013
National Library of Scotland: town plans. http://maps.nls.uk/towns/. Accessed 8 May 2013
National Library of Scotland: OS abbreviations. http://maps.nls.uk/os/abbrev/. Accessed 8 May 2013
Nes A van, Lopez MJJ (2007) Micro scale spatial relationships in urban studies: the relationship between private and public space and its impact on street life. In: Proceedings of the 6th international space syntax symposium, İstanbul, 023

Nes A van (in preparation) Space syntax in urban studies: an introduction, Sage
Nijman J (2007) Introduction: comparative urbanism. Urb Geogr 28(1):1–6
Noizet H, Costa L (2013) Alpage: AnaLyse diachronique de l'espace urbain PArisien: approche GEomatique. http://alpage.tge-adonis.fr/en/. Accessed 6 May 2013
Noizet H, Grosso E (2011) The ALPAGE project: Paris and its suburban area at the intersection of history and geography (9–19th century), In: Proceedings of the 25th international cartographic conference (ICC), 3–8 July 2011, Paris (France). http://halshs.archives-ouvertes.fr/docs/00/66/84/00/PDF/aci_2011_ALPAGE_project.pdf. Accessed 6 May 2013
Oliver R (1993) Ordnance survey maps: a concise guide for historians. The Charles Close Society, London
Palaiologou G, Vaughan L (2012) Urban Rhythms: Historic housing and sociospatial boundaries. In: Greene M, Reyes J, Castro A (eds) Proceedings of the 8th international space syntax symposium, Santiago de Chile, 8161
Paliou E, Corsi C, Vermeulen F (2012) The whole is more than the sum of its parts: geospatial data integration and analysis at the Roman site of Ammaia (Marvão, Portugal). Presented at: 40th computer application and quantitative methods in archaeology conference, Southampton
Paliou E, Knight DJ (2010) Mapping the senses: perceptual and social aspects of late antique liturgy in San Vitale, Ravenna. In: Proceedings of the 38th computer application and quantitative methods in archaeology conference, Granada. http://www.academia.edu/2480702/Paliou_E._and_D.J._Knight_in_press_._Mapping_the_senses_Perceptual_and_Social_aspects_of_Late_Antique_Liturgy_in_San_Vitale_Ravenna_._Proceedings_of_CAA_2010_Computer_Applications_and_Quantitative_methods_in_Archaeology_International_Conference_Granada_6-9_April_2010. Accessed 6 May 2013
Palliser DM (ed) (2000) The Cambridge urban history of Britain, vol I. Cambridge University Press, Cambridge, pp 600–1540
Pinho P, Oliviera V (2009) Different approaches to urban form. J Urb: Int Res Placemaking Urb Sustain 2(2):103–125
Poehler E, Pompeii bibliography and mapping project. http://digitalhumanities.umass.edu/pbmp/. Accessed 6 May 2013
Pratt AC (1995) Putting critical realism to work: the practical implications for geographical research. Prog Hum Geogr 19(1):61–74
Rice DS, Culbert TP (1990) Historical contexts for population reconstruction in the Maya lowlands. In: Culbert TP, Rice DS (eds) Precolumbian population history in the Maya lowlands. University of New Mexico Press, Albuquerque, pp 1–36
Robinson J (2004) In the track of comparative urbanism: difference, urban modernity and the primitive. Urb Geogr 25(8):709–723
Rodier X, Saligny L, Lefebvre B, Pouliot J (2009) ToToPI (Topographie de Tours Pré-Industriel), a GIS for understanding urban dynamics based on the OH_FET model (Social Use, Space and Time). In: Proceedings of the 37th computer application and quantitative methods in archaeology conference, Williamsburg
Sanders WT, Webster D (1988) The mesoamerican urban tradition. Am Anthropol, New Ser 90(3):521–546
Sayer A (1981) Abstraction: a realist interpretation. Radical Philos 28:6–15
Sayer A (2000) Realism and social science. Sage, London
Scobie GD, Zant JM, Whinney R (1991) The brooks, winchester: a preliminary report on the excavations, 1987–1988, Archaeology report 1. Winchester Museums Service, Winchester
Smith B (2001) Fiat objects. Topoi 20(2):131–148
Smith B, Varzi AC (1997) Fiat and bona fide boundaries: towards an ontology of spatially extended objects. In: Hirtle S, Frank AU (eds) Spatial information theory a theoretical basis for GIS. Springer, Berlin, pp 103–119
Smith ME (1989) Cities, towns, and urbanism: comment on sanders and webster. Am Anthropol, New Ser 91(20):454–460
Smith ME (2003) Can we read cosmology in ancient maya city plans? comment on Ashmore and Sabloff. Lat Am Antiq 14(2):221–228

Smith ME (2007) Form and meaning in the earliest cities: a new approach to ancient urban planning. J Plan Hist 6(1):3–47
Smith ME (2011a) Empirical urban theory for archaeologists. J Archaeol Methods Theor 18:167–192
Smith ME (2011b) Classic maya settlement clusters as urban neighbourhoods: a comparative perspective on low-density urbanism. J Soc Am 97(1):51–73
Smith ME (2012) The role of ancient cities in research on contemporary urbanization. UGEC Viewpoints 8:15–19
Smith ME, Peregrine P (2012) Approaches to comparative analysis in archaeology. In: Smith ME (ed) The comparative archaeology of complex societies. Cambridge University Press, New York, pp 4–20
Smith ML (2008) Urban empty spaces: contentious places for consensus-building. Archaeol Dialogues 15(2):216–231
Stanley BW, Stark BL, Johnston KL, Smith ME (2012) Urban open spaces in historical perspective: a transdisciplinary typology and analysis. Urb Geogr 33(8):1089–1117
Stöger J (2011) Rethinking ostia: a spatial enquiry into the urban society of Rome's imperial port-town. Leiden University Press, Leiden
Taylor P (2012) Extraordinary cities: Early 'city-ness' and the origins of agriculture and states. Int J Urban Reg Res 36(3):415–447
Thaler U (2005) Narrative and syntax: new perspectives on the Late Bronze Age palace of Pylos, Greece. In: van Nes A (ed) Proceedings of the 5th space syntax symposium. Techne Press, Amsterdam, pp 323–339
Tourtellot G III (1988) Excavations at Seibal, Department of Peten, Guatemala: Peripheral survey and excavation, settlement and community patterns, vol 16, Part 1, Peabody museum of archaeology and ethnology, Harvard University, Cambridge, Mass
Trancik R (1986) Finding lost space: theories of urban design. Wiley, Hoboken
Valente V (2012) Space syntax and urban form: the case of late medieval Padua. Post-Class Archaeol 2:147–166
Vis BN (2009) Built environments, constructed societies: inverting spatial analysis. Sidestone Press, Leiden
Vis BN (2013a) Establishing boundaries: a conceptualisation for the comparative social study of built environment configurations. Sp Flows: Int J Urb ExtraUrb Stud 2(4):15–29
Vis BN (2013b) Boundary concepts for studying the built environment: a framework of socio-spatial reasoning for identifying and operationalising comparative analytical units in GIS. In: Proceedings of the 40th computer application and quantitative methods in archaeology conference, Southampton
Vis BN (forthcoming) Boundaries of the built environment: a key to the comparative understanding of the everyday structure of settled societies. J Borderland Stud (under revision)
Wallace S (2011) Contradictions of archaeological theory: engaging critical realism and archaeological theory. Routledge, Oxford
Ward K (2010) Towards a relational comparative approach to the study of cities. Prog Hum Geogr 34(4):471–487
Whitehand JWR (1981) Background to the urban morphogenetic tradition. In: Whitehand JWR (ed) The urban landscape: historical development and management. Academic Press, London, pp 1–24
Whitehand JWR (2007) Conzenian urban morphology and urban landscapes. In: Proceedings of the 6th international space syntax symposium, İstanbul, II
Wood D (1992) The power of maps. The Guilford Press, New York
Wunderlich AL, Hatcher RD Jr (2009) Rescuing legacy digital data: maps stored in adobe illustrator™ format. In: Soller DR (ed) Digital mapping techniques 08—Workshop proceedings, U.S. geological survey open-file report 2009, 1298. http://pubs.usgs.gov/of/2009/1298/. Accessed 8 May 2013

Yeung HW (1997) Critical realism and realist research in human geography: a method or a philosophy in search of a method? Prog Hum Geogr 21(1):51–74

Yoffee N (2009) Making ancient cities plausible. Rev Anthropol 38:264–289

York AM, Smith ME, Stanley BW, Stark BL, Novic J, Harlan SL, Cowgill GL, Boone CG (2011) Ethnic and class clustering through the ages: a transdisciplinary approach to urban neighbourhood social patterns. Urb Stud 48(11):2399–2415

Part II
Mapping Cities: Lyon and Barcelona as Case Studies

Lyons, the Spatial Analysis of a City in the 17th and 18th Centuries. Locating and Crossing Data in a GIS Built from Written Sources

Bernard Gauthiez and Olivier Zeller

Abstract The Lyons historical GIS was developed from the end of the 1990s. Its aim was to reach a new understanding of the transformation of urban and social spaces by spatializing data at the buildings' scale. We thought that by such a jump in precision of a factor 100, from a subdivision by 36 quarters to one by 3,500 buildings in the modern period, the mapping would lead to new perspectives and results in urban history. This involved working through two centuries of numerous and heavy archival records, taxes registers, censuses, building permits, property changes…, each record comprising generally between 3,500 and 8,000 entries, in order to create critically researched data bases followed by vector GIS layers. This required developing a method to reveal the implicit spatiality of these written sources, and to establish a mapping topography, allowed by a careful and geometrically checked reconstruction of the city's plot pattern together with its variation before 1800, with the help of numerous archival maps. It was also necessary to take into account the space transformation, studied at the actual scale of individual investments, owing to the building permits checked with the still extant constructions, and the administrative record. The changing built-up and social spaces are a constant preoccupation in our work. In short, understanding this process of change is absolutely necessary to analyze a situation at a given date, and its significance at different time scales in itself. This paper describes the way we achieved the intended goal, working together as a geographer who was also an historian, and an historian who became geographer. It develops some results and draws the research lines on which the work is in progress, or still to be done. One very interesting aspect of the method is that, once the historical address system has

B. Gauthiez (✉)
Centre National de la Recherche Scientifique, UMR 5600 Environnement, Ville, Société, Université de Lyon (Lyon 3), 5 rue Ste-Catherine 69006 Lyon-France, Lyon, France
e-mail: bernard.gauthiez@gmail.com

O. Zeller
Centre National de la Recherche Scientifique, UMR 5600 Environnement, Ville, Société, Université de Lyon (Lyon 2), UMR 5600 Environnement, Ville, Société, Lyon, France
e-mail: pr.olivier.zeller@free.fr

S. Rau and E. Schönherr (eds.), *Mapping Spatial Relations, Their Perceptions and Dynamics*, Lecture Notes in Geoinformation and Cartography,
DOI: 10.1007/978-3-319-00993-3_5, © Springer International Publishing Switzerland 2014

been established, in attributing successive owners to the same building, many sources become 'spatializable', including non-serial ones, and so deliver new historical insights.

1 Introduction

The Lyons historical GIS was conceived in the late 1990s in order to use some still unexploited archival sources, notably the building permits given by the city administration, the Consulat, from 1600 to 1763 (Gauthiez and Zeller 2006, 2009). This source constitutes a continuous series of about 8,000 acts, including about 2,535 complete reconstructions or new constructions and 840 raisings of either one or two floors. The rest are façade modifications, placing or removing of windows and doors, and water mains authorizations, together with urban signs.[1] This project had a primary goal: to map the evolution of the city at the scale of its elementary elements, in other words the 'quanta' of modifications, i.e. each new building, new story added, façade modification, and to see what this mapping could add to the understanding of the transforming city space. This transformation is not continuous, but is due to the multiplication of these tiny changes, in some cases according to relatively large schemes and planned developments whereas the identification of these schemes is another matter for investigation.

Another goal, only possible if the preliminaries for the first one were fulfilled, was to construct a cartography of the city on the base of each building or plot.

This project was extremely challenging as no vector plan of the pre-1800 period existed, and as the building permits were given only according to a proprietor, and indicated only a street location, without a proper address system, which was not developed until 1790 at Lyons. So it was necessary to reconstruct a precise topography of the city, and to try to build an address system from written records, based on the successive proprietors.

The project has by now been largely implemented, as about 2,280 (90 % of 2,535) of the new constructions and 680 (80 % of 840) of the raisings have been accurately mapped. But, to reach the initial goal, it was necessary to develop a completely new method, or, to say the least, to explain it, as some hand-working erudite scholars had already followed on along that path (Boudon et al. 1977; Claval and Claval 1986). Working in such a way, new, often unexpected fields were opened, which we are still trying to investigate thoroughly.

[1] Lyons Municipal Archive (hence LMA), DD series registers.

2 Building the GIS

2.1 Topographical Basis

To build the topographical basis, the following method has been developed (Schätti and Viaccoz de Noyers 2005; Gauthiez 2008; Gregory 2003; Gregory and Ell 2007):

First step: drawing a vector plan of the city based on the first cadastral plan of 1830–31, correcting its geometry in order to make the same objects correspond exactly with their present locations (georeferencing and rubber-sheathing are clearly not the right method). Three main layers were drawn: plots, buildings, other topographical features such as the bridges, river banks, stairs, etc.

Second step: based on this plan, and using more than 300 partial 18th century maps of the city,[2] such as seigneury plans, a vector plan of Lyons c. 1745 was established on the same topographical basis, also in three layers.[3]

2.2 Address System

The sources allowed for only one possible comprehensive address system, i.e. the mapping of the successive proprietors of the buildings, as this record is the only more or less complete one that can be found in the archives. We were lucky as housing in Lyons was already largely pluri-familial in the 16th century, and not mono-familial as was generally the case in northern Europe, so we had to deal with a relatively small number of buildings, about 3,500.[4] Two complementary methods have been used to achieve success with the mapping process:

[2] For example LMA 2 S 26, *Atlas de la rente noble de l'archevêché*; Archives Départementales du Rhône (hence ADR) 10 G 2338, St-Jean chapter; ADR 15 G 188, St-Nizier chapter.

[3] This plan was first drawn by hand in the early 1990s (Gauthiez (1993) Lyon, formation et évolution d'un espace urbain, 1.- cartographie du site et Moyen-Âge. Ecole Nationale Supérieure d'Architecture de Lyon, Vaulx-en-Velin). It was vectorized and its geometry corrected in the mid 2000s.

[4] This peculiar aspect of the building pattern in Lyons has to be explained. It developed in a first phase in the early 16th century, and became dominant from the 1640s. It implies that average financing units emerged from a capitalistic development, and a dwelling pattern adapted to the low wages of the silk industry, on which the city's wealth was based, in fact the wealth was confined to only a small part of the population, notably the *marchands-fabricants*. In contrast, the average early 17th century building looks like raised houses, which they often were. In the course of time, the proportion of low incomes passed 80 % and the wealth became concentrated (Garden 1970), as a consequence, whilst the average building grew in size, the average number of new constructions decreased accordingly. The concentration of real estate and economic wealth was extreme in 1789. For Paris, see Cabestan 1997.

As a consequence of the growing size of the new buildings, their increased number of floors up to seven, and of the grouping of 2–3 previous plots to build them, the built-up area at Lyons expanded by about 30 % between 1600 and 1789, although the population had more than tripled, from c. 40,000 to c. 150,000 inhabitants.

- From the 1830–31 cadastre, and using the numeration system (even and odd numbers) established c. 1810, it was possible to map the proprietors in 1814–15, known from this year's census.[5] The first complete census was in 1808, but the numeration system was different at that time (this system was in use from 1790 to c. 1810). As the owners are very often the same in 1808 as in 1814, it was possible to map 1808 by keeping the names where they were documented at both dates, and by filling the gaps from the 1808 census, whose houses are numbered in numerical order (1, 2, 3… 148, for example) for each quarter.[6]
- The second phase involved the use of the listing order of the older tax documents to reconstruct the space order of the proprieties in the actual built up streets. It has been possible to prove that this registering order had been quite stable through the 18th century, even from the mid 17th, with only limited variations of the tax agent's route (Gauthiez and Zeller 2010). When compared to the 1808 census made according to numerated buildings, and as some 30 % of the proprietors are the same, the route and proprietors in 1789 given in the 20th (*Vingtième*) tax list can be reconstituted with a fair degree of certainty.[7] The 20th tax list in 1766 can be mapped rather easily thanks to a very high proportion of common proprietors compared to 1789, and so on. Unfortunately, the series of remaining sources (the identified ones, as some unexplored archives may reveal new ones) make it impossible to go earlier through time when using this tax record, but other sources can be very useful for earlier dates. The lantern tax list of 1698 has been mapped by using the *Règlement Général d'Alignement* of 1680,[8] a text giving street by street the line of the façades to be respected if a building were to be reconstructed, and which mentioned nearly all the angle houses proprietors, i.e. about 800.[9] From this source and the unfortunately partial list established in 1677 for a city loan, it is possible to redraw the sequences of the route of the tax agent.[10]

---

[5] LMA 921 WP 023-025.

[6] LMA 921 WP 005-010. The numeration system established in 1790 is an evolution of the previous one, determining a lesser number of quarters and attributing the number to a given building, rather than to the order place of a proprietor.

[7] The 20th tax is annual and based on 1/20th of the rent value of a property, generally a single building, but sometimes a group, or a part of a single building. As the annual rent value is about 1/20th of the real estate value of the building, the 20th tax amount equals more or less 1/400th of the total value of the building (or group of them, or part of it).

[8] LMA DD 56.

[9] The *Consulat* decided in 1698 to establish a public street lighting system, which was financed by a tax based on the building's value, LMA FF 0754-0755. The 1680 *Règlement Général d'alignement* is a by-law adopted in December 1680, to determine for all streets and squares the façade lines each new building had to respect. Its goal was to regularize the street width in a context of rapidly growing traffic, and to organize the embellishment of the city.

[10] The *Consulat* went bankrupt in 1677. Among various measures, it was decided to impose a loan on all the rich people in the city. An inquiry followed in every quarter, but half of the results are unfortunately lacking, LMA CC 4187. The mapping of the wealth in the city, despite being partial, is quite interesting.

These reconstructed proprietors' locations, fairly reliable for 1814, 1808, 1789,[11] 1766,[12] 1698[13] and 1677 (when the data is available),[14] can be extended, but only partially, in conjunction with other partial sources, for example 1744, 1723, 1636. The system can be in fact strengthened by the location of the building permits. Once a permit can be located according to the name of the proprietor involved, its neighbors, generally mentioned in the deed according to North, South, East and West, can be also easily located, and so give some new proprietors' names. This is of great importance as more than 50 % of the rebuildings and façade modifications were carried out by a series of neighbors.

These locations can be matched to the still extant 17th and 18th century buildings, one third of the constructions erected, and documented in another vector layer in the GIS.[15] The iconography, despite being rather thin, can also be of some help.[16]

Each source forms a particular layer which can be matched with the other ones to cross data. The description and approximate dating of the buildings using art history methods, including the survey of urban signs and of the proprietor's initials which are often placed over the main door, constitutes another very useful layer.

Years of mapping and crossing the available information from many sources have led to a rather secure system of location of the proprietors, and thus of a secure location of the building permits. The certainty is more or less total for the central streets, which are more documented, and where many buildings are still extant. The locations in the areas that have disappeared and along the longer streets (Grande Côte, Pierre-Scize street running more than 1 km) where the sequences of proprietors may shift, retain generally some uncertainty, for example a location can be displaced to the neighboring plot according to new information. It seems that the city was administered quite conscientiously, as the building permits were, with only a few exceptions, systematically followed on by the actual transformation or construction of a building, and those who tried to escape the authorization, or simply ignored its necessity, were fined. Some had even to demolish what had been erected without authorization, or to rebuild according to a different façade line, which was indeed quite expensive. Few areas are a little less informed, mainly the St-Jean Cathedral area, where the jurisdiction of the chapter made the canons master of the building permits, and for which the record is not complete, and the area of the fortifications where the military administration had the jurisdiction, whose record seems to be lost.

---

[11] LMA 3 C 52.

[12] LMA 2 II 160.

[13] LMA FF 0754-0755.

[14] LMA CC 4187.

[15] Many still extant buildings also bear the initials of the owner's name, placed in wrought iron over the main door. Some rare dates are still in existence. This was made by undertaking a field survey.

[16] But it tends to nearly always represent the same places. And the iconography is somewhat rare at Lyons.

From this rather burdensome work, which was only made possible by searching through many important archival sources with each one set up in a database and then a vector layer associated to a database,[17] a new world of urban history comes alight.

Here are some examples of the new horizons now opened.

2.3 The Routes of the Administrative and Tax Agents

This reveals how space was surveyed when no plans were used, according to various route determinations. The way each one of the 36 *quartiers* (or *pennons*) of the city was surveyed is very varied (Gauthiez and Zeller 2010)[18] (Figs. 1 and 2). This implies that each quarter administration had to determine its own way to list the proprietors at some moment in time (probably in the early 16th century). The route followed by the tax agent was very stable and subject only to small modifications. This became also the base for the progressive introduction of a numerating system from 1723. But, if the route was exactly the same in 1766 and 1789, the number of a given building was shifting as it was first of all the number of a given property in a list.

The reconstruction of the agents' routes allowed for a precise mapping of the quarters as each property is part of a given one, and its evolution (Fig. 3). So it opens a new window on the administration's history. This allows for all the mapping subsequently evoked.

2.4 Mapping the Building Permits

As these were made according to objects which were the results of retraceable social processes (Gauthiez 2004), this enables the reconstruction of the film of the urban fabric transformation and leads to questioning its logics: location, social milieus, and uses when crossed with economic data. The film of the building investment highlights the periods and places of higher and lower intensity. It also discloses some scale effects previously not referred to, especially the way the new buildings follow a 'chain' pattern, i.e. very often. When one was replaced, the neighboring houses were also rebuilt, or at least renewed in the subsequent years.

[17] The obvious method would be to make a 'juncture' between a layer of polygons and a database, but, as the lines in the tax database do not systematically correspond to a single building, but sometimes also to a part of one or a group of them, each line has to be reexamined. For instance, the 1789 20th tax documents list 3,821 entries, but only 3,431 buildings existed in 1786 (Brac 1787).

[18] The number of pennons varied slightly through time, but the central ones remained constant. Their limits also varied slightly.

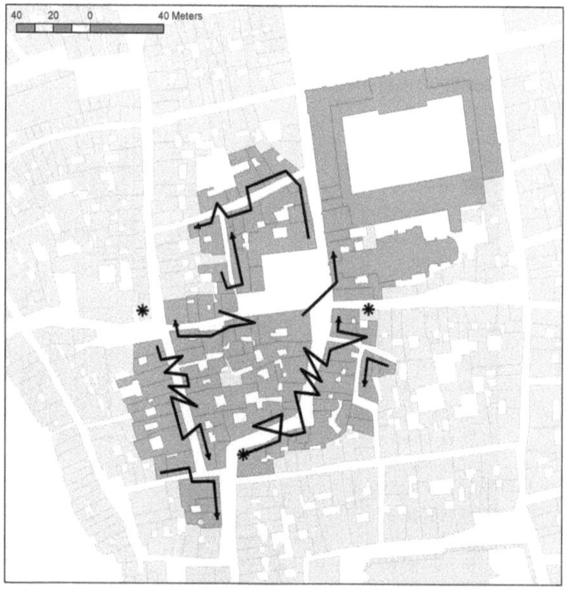

Fig. 1 Route of the administrative agent, Lyons, *quartier* St-Pierre, 1677. A case of a complicated route, explained by the passing from door to door between the two sides of a same street (B. Gauthiez©2013)

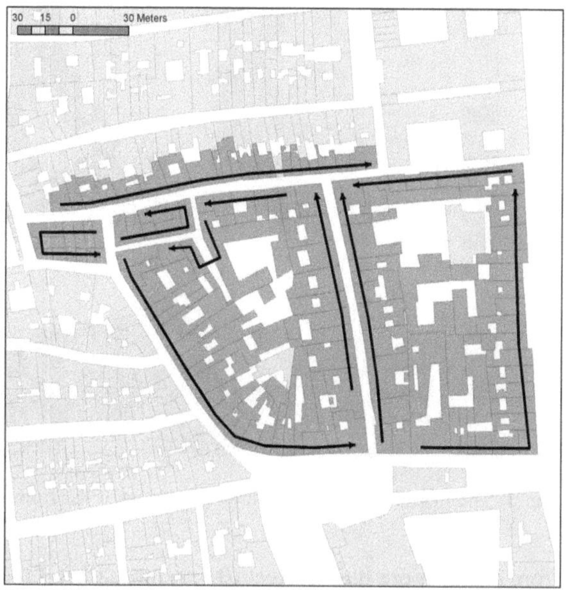

Fig. 2 Route of the administrative agent, Lyons, *quartier* Cordeliers, 1677. An example of a rationalized route, examining the properties going around the same block, one at a time (B. Gauthiez©2013)

This process was certainly encouraged by the Consulat administration, as many places affected by this process were streets to be widened and/or embellished. This process concerns probably more than 50 % of the new constructions. Another effect of the rebuildings was made obvious by mapping the widening of the streets on account of the obligation to place the new façade backwards, in order to gain

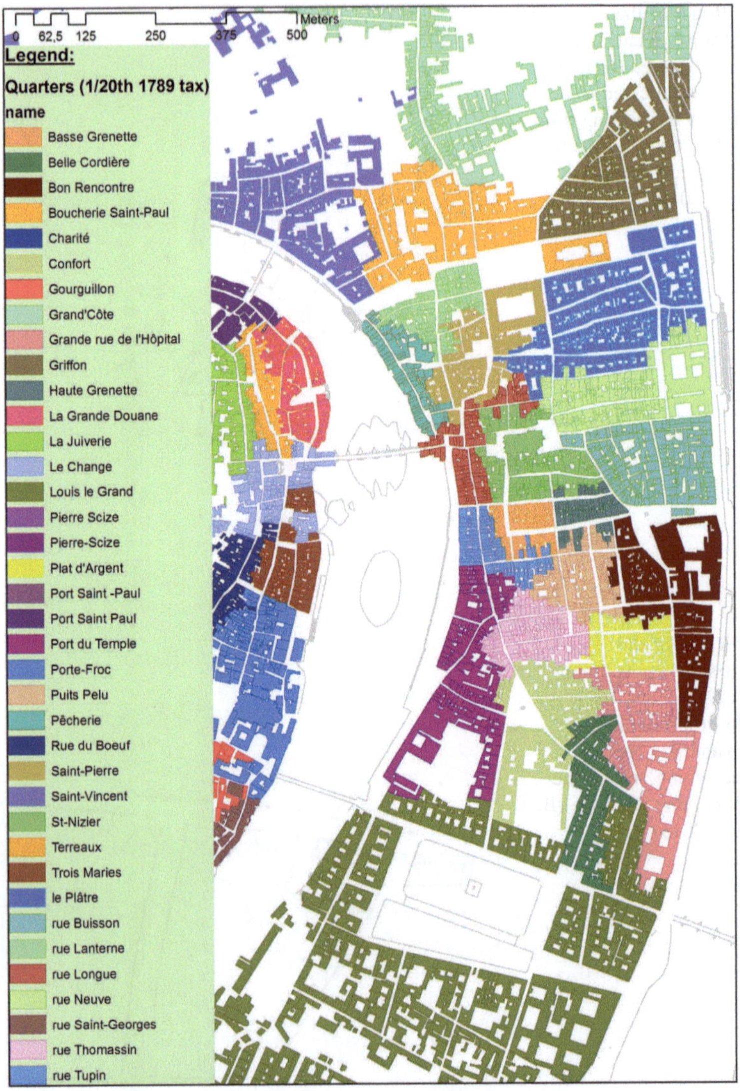

Fig. 3 The map of the Lyons central quarters in 1789, from the 20th tax (B. Gauthiez©2013)

from one to more than 10 feet (0.3 to more than 3 m) of street width (Fig. 4). The Consulat street policy was constant and very efficient in that respect. The general street widening was in response to the increasing traffic of coaches and carriages, and was also intended to ease the traffic jams. This change took place in the early 17th century and became common from the 1660 s. It probably signifies a great change in the way commercial goods were carried, using previously principally mules and horses then more and more carriages, including heavy ones. It is

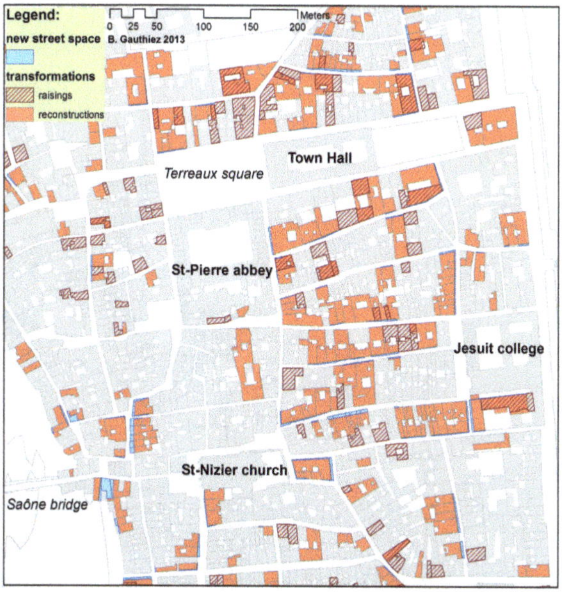

Fig. 4 Cartography of all the constructions and raisings in the centre of Lyons between 1698 and 1763 (match with Figs. 6 and 7) (B. Gauthiez©2013). The map shows in *red* a concentration of new constructions (the raisings are hatched) between St-Nizier church and the Hôtel de Ville, where the land value grew steadily during this period. This concentration contrasts with a sparser renewal elsewhere in the area. The *blue* spaces indicate the street widening operated during the same period and the opening of the quay on the Saône bridge extremity in 1719

probably a major step towards the Industrial Revolution. Lyons developed very early as a great industrial city. This change had not previously been noticed.

This mapping is clearly very efficient for determining and interpreting the planning policy of the municipality (Gauthiez and Zeller 2013).

2.5 *Location of the Weaving Looms in 1744 and 1833*

From the 17th century Lyons, a great industrial city whose economy was based on the silk industry, experienced several social upheavals, the most important one in the modern period being the 1744 revolt (Zeller 1990). This revolt led to several decisions of a mainly political nature. The police quarters' division was reorganized in 1746,[19] the Consulat was also reorganized, and a register of the masters and workers of the silk industry (the 'Grande fabrique') was established in 1745.[20] Its main scope was to gain a better control of this shifting rowdy population, a goal

[19] LMA EE 11, LMA 3 S 693.

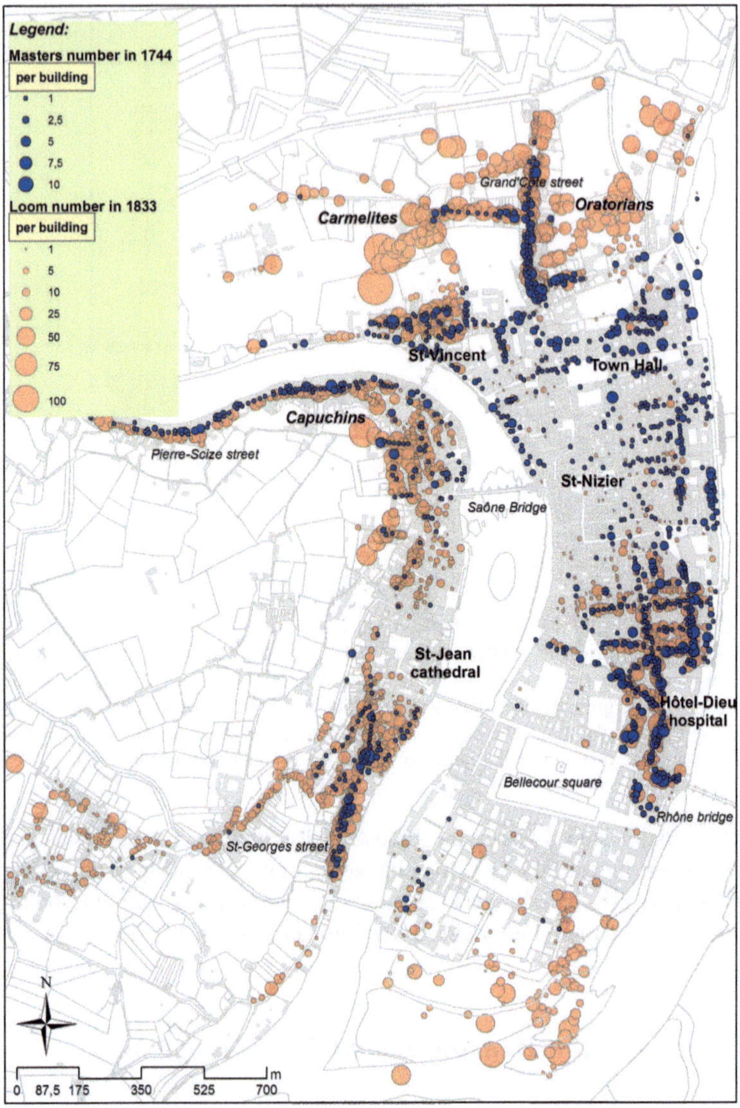

Fig. 5 Cartography of the silk masters in 1745 and of the silk weaving looms in 1833, by building, at Lyons (B. Gauthiez©2013)

probably only partly achieved, as no police system ensued. We mapped the location of the silk masters in order to have a better understanding of the geography of the silk industry in the city (Fig. 5). The image is striking on two different levels: the masters (and the workers, one worker being generally employed by a

[20] LMA HH 578.

master) were generally located in the outskirts of the city, but still present not far from the center (in blue). In the center and the aristocratic areas such as Place Bellecour, neither workers nor loom were present. They were concentrated in the poorest sectors of the city. The 1677 data gives a clue on the progressive displacement of the silk industry from the center, which the map of the weaving looms in 1833 later confirms. This is a *longue durée* trend lasting roughly two centuries, and only clearly apparent on account of mapping. Another unexpected scale is the frequent grouping of the silk masters in particular buildings. This implies that a specialization took place in some buildings, all the more apparent when one crosses the map with the building authorizations. Some buildings were obviously financed and erected to house silk weaving looms and their workers. As no typological adaptation of the buildings appears, this is also only visible thanks to mapping. The way some private investors and the Hôtel-Dieu hospital built houses for silk weaving has still to be investigated. The general view in Lyons's history is that each silk master was working independently, possessing their own weaving looms and operating in their apartments. It is obvious now that the regrouping of the masters was organized and that some buildings were functioning more or less like manufactures, even if their occupants were not wage-earners, and even if they were said free artisans.

This was all the more obvious in 1833, the year of a census registering the loom number by building.[21] The information was needed, as in 1745, for another revolt had occurred in 1831, one of the two famous 'Canuts' revolts (*canut* = a silk worker, in fact also to the many various jobs related to this industry. The second *canuts* revolt took place in 1834). The city center was then also held for some time by the *canuts*. So, in two similar circumstances, the administration reacted, at 90 years' distance, in a same way, notably by establishing a census, precious for us as its provides previously unknown information.

Between 1745 and 1833, and overall after 1810, the silk industry had spread considerably, following directions already perceptible in 1745: with old linear suburbs to the west, at St-Georges and to the north, and around the Hôtel-Dieu hospital. It had also acquired new areas around the city, to the north, in new developments in religious tenements suppressed during the Revolution such as the Oratorians and the Carmelites where the main concentration of looms was situated in 1833. The same process was also operating at the Capuchins, to the west. On this side of the city, many buildings previously occupied by rich people, merchants and bankers, were now occupied by poor people weaving silk. The balance between the two sides of the Saône river changed, with the west side pauperizing and the east side gentrifying. The tendency to produce adapted buildings to place weaving looms is still clearer in 1833, as some new developments are fully dedicated to the silk industry. For example in Blanc Street, established through the Carmelites convent's close, all the new buildings, erected from 1820, were occupied by silk weavers. In one of them 139 looms were counted, just a little less

[21] LMA 921 WP 151-158.

than in the nearby "immeuble aux 400 fenêtres" ("400 windows building"), with 158 looms. At the Capuchins, the old convent was sheltering some 119 of them. The expression *caserne ouvrière* (workers' barracks) was sometimes used to describe these concentrations making the buildings appear like manufactures or mills. As the grouping of the silk workers was organized and through the rents quite profitable to investors (with some money coming from silk trade), the official discourse referring to the silk workers as free artisans, and the folklore built up accordingly and still today embellishing their very hard life, seem rather cynical. In 1834, a new revolt exploded, with a more political tone; the city was again occupied and managed by the workers, about 1,000 workers and soldiers had died when order was reestablished by the government.

The industrial development, putting aside some central pauperizing, was mainly an outer fringe phenomenon, but some rich areas also developed at the periphery with the St-Clair quarter at the north-east, and around the old Ainay abbey and Bellecour royal square to the South.

3 An Example of Complex Data Processing: The Land Value in 1698, 1766 and 1840

Establishing the land values is extremely rich in information about the link between social and built up space, mainly in relation with the age of the building and its location pattern, which may reflect economic and/or social polarities.

Its mapping represents therefore a significant breakthrough for our knowledge of the city's history. To achieve such a scope, the method has followed this path:

First:

Establish the reality of the built up area at the particular periods of time which correspond to the available serial sources. This has been described in the previous part of this paper.

Second:

Calculate the value of the buildings from the sources. This is rather easy, as, the sources being mainly taxes, their amounts are derived from the rent value of the buildings. This means actually calculating the annual income perceived by the owner from the tenants and establishing the tax as a proportion of this income. As the annual income is generally considered as about 1/20th of the building's value, the amount of the 1/20th 1766 tax, for example, equals 1/400th of this value. In 1698, the lantern tax amount is 1/1,000th of the building's value, and, in 1840, the municipal census gives the rents of all the dwellings and activities included, which, when totaled, is about 1/20th of the total value. In this way, it is possible to establish the values of each property, according to consistent valuation in the rent market, in 1698, 1766 and 1840, fortunately at some fairly equal intervals (68 and 74 years). Our investigations in the archival records have given, at the moment, no other source providing this value for the period, except for the years 1789 and 1790.

Third:

Calculate the total floor area of the buildings. This is a task involving more incertitude and difficulty. The area of the buildings is ascertainable for the 1830 cadastre maps and earlier documents when available and re-drawn as has been previously described. This area has to be checked according to the reality of existence of the building at the tax's date, which we know from the reconstructions carried out at a later date and for still extant buildings. We know all the buildings extant in 1840, described in the census and very close to the 1830–31 cadastre,[22] but, when we go backwards in time, the number of buildings for which we lack sufficiently precise information increases accordingly. The land value has been calculated as a floor value, i.e. a value per m^2, because the height of the building may vary from three to seven stories (or 2–6 floors, excluding the ground floor). We have excluded the cellar story(ies). The land value calculated in that way is significant as a comparison value (which does not actually exclude the significance of a building's value estimated as a whole, which is interesting when considering the estate market and the rebuilding logic). The storey number can be inferred from the 1840 census and is sometimes mentioned in the 18th century maps. It had been corrected by using the building permits, when they are authorizations for adding one or two stories.

In such a way, notwithstanding the incertitude due to uncertainties about the actual height and existence of the relevant buildings and calculating the areas including walls and stairs, it has been possible to map the floor values for ¼ of the city of Lyons in 1698, 1766 and 1840. The result produces information about the estate values polarities at a given date, and their dynamic evolution through about one and a half centuries. In the three maps, all the values have been translated into the 1766 *livre* value in order to correct the bias due to the inflation.

The 1698 values (Fig. 6) show a clear polarity of the highest values around the eastern extremity of the Saône river bridge, an ancient situation, as this bridge had long been the only one crossing the Saône at Lyons. The sector of Mercière Street, to the south of the bridge, was also of high value. This street led to the bridge crossing the Rhône river to the south eastern part of France. The great difference between the groups of lowest and highest estate values, more than four times in a space of less than two hundred meters, is very striking. The 1766 values (Fig. 7) reproduce the same pattern, but a new area of high value appears between St-Nizier church and the town hall (Hôtel de Ville), more centrally situated, along another street (Clermont Street), and clearly associated to a large renewal of the built up fabric of the quarter, as shown on Fig. 4. The highest values are also linked to the street widening, particularly around the extremity of the Saône river bridge and the streets running West-East at the center of the figure (Mulet, du Bat-d'Argent, de l'Arbre-Sec, Pizay streets). The 1840 map (Fig. 8) shows a strong

[22] The exact number of floors can be difficult to define precisely, as the roof can be partly used. This was increasingly the case through the 18th and 19th centuries, with the addition of roof windows. It was generally a space not used previously, as the roof slope is low.

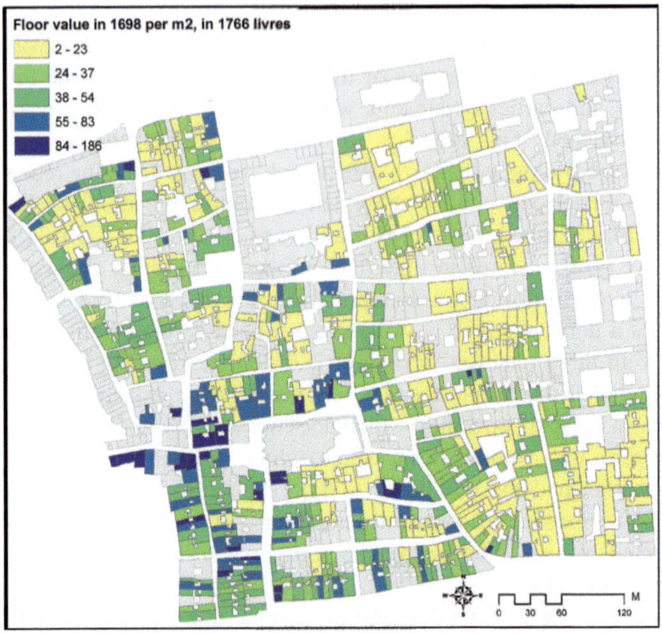

Fig. 6 Floor m^2 values in the center of Lyons in 1698 (converted into 1766 *livres*) (B. Gauthiez©2013)

Fig. 7 Floor m^2 values in the center of Lyons in 1766 (into *livres*) (B. Gauthiez©2013)

Fig. 8 Floor m^2 values in the center of Lyons in 1840 (converted into 1766 *livres*) (B. Gauthiez©2013)

evolution, although the built-up fabric had undergone little change. The middle values are more diffused, but the highest ones are neatly concentrated on a few main streets, one from des Terreaux square to the Saône bridge and southwards to the quay opened in 1719, two, to a lesser degree, situated around the town hall and Clermont Street, which had been already developing. The Grenette quarter (low centre of the figure) was losing some of its value, which means of course some of its former economic importance. One can clearly see the importance of the traffic and of dealing with it by widening the streets and creating improved links. The high values, however, are concentrated in only a few streets.

One should be prudent about the social significance of these values, taken as a whole for each building. Actually, the highest dwelling values are not superimposed on the highest activity values, which can be differentiated only in 1840 (map not shown), thanks to the census, and could be also mapped from the 1790 data used in Gardens's study, but only, at that time, at a non significant scale. As a consequence these maps show mainly the concentration of the high value activities, notably the luxury trades, workshops, and shops. The 1840 census indicates that the new buildings between the town hall and St-Nizier church were principally housing upper class people, because of their large apartments with high rent values, but a value per m^2 not so high on average.

4 Perspectives on Urban History

4.1 Rhythms, Spaces and the Makers of Built-Up Urban Fabric and New Developments

Placing the building permits according to the successive proprietors and the informed buildings (still extant or thanks to iconography) makes it possible to identify the more dynamic areas, as an object of gentrification or renewal, and the stagnant ones, growing older because they were not replaced and became lower in value and more pauperized for that reason (Gauthiez 2004). An important transformation can have a big induction effect on their direct neighbors and on their area. This can be due to a public or a religious building, but also to a private one. Some of these looked palatial and were undoubtedly impressive. Investors could act according to familial property holdings logic, by applying micro-scale speculation. Some investors, especially builders and architects, often acted as entrepreneurs, already in the mid-17th century, a process more present when they developed, with the aid of the companies they had formed, several new areas in the 18th century (Zeller 1995; Gauthiez 1995). But financial aspects were not the only ones, as technical and esthetical aspects of *commoditas* greatly improved and changed in the period, and as symbolical and social status were also factors of transformation, be it by fashion.

4.2 Mapping Rent Values

The *Contrôle des actes* (public register of notarial deeds), an archival record carried from the early 18th century, enables the rent values to be mapped for many buildings kept under the regime of *location principale*, which means that a main tenant contracted with the proprietor for the whole building and then sub-contracted with the sub-tenants of the apartments and shops. The total of these secondary rents was generally superior to the contracted primary rent when the main tenant was successful, but sometimes inferior and thus to his detriment. For the previous period, one has to work through partial surveys in the notarial records. The rent variables, age and repair state of the building, comfort level, size and location can then be studied. Information on the social statuses of the dwellers is of great importance. The changing relation between masters and workers, from lodging at one's master's home to owning an apartment accommodating wife and children, contributed to the development of socially segregated areas, especially where the workers could find low rents, in former increasingly derelict suburbs (Pierre-Scize, St-Georges and Grand'Côte streets for example at Lyons). On the *longue durée*, the general trend from 16th to 19th century is that owners used to live in their property, then in a neighboring one, then farther and farther away, so the control of the tenants became more and more financial and abstract.

4.3 Mapping Social Space Fundamental Variables

Some other perspectives have been pursued: wealth levels, geographical origins, profession specializations, religious minorities, women-led families (especially widows)... But the question of male and female habitats has to be more closely examined, bearing in mind their life course in space (even the vertical space of the stories). It was at least partly socially and professionally conditioned. Mapping at the building scale, health professions, inns and pubs, dangerous, noisy, unre-spectable activities, assistance places, will be certainly quite significant, which is also true for control points for commerce and people coming in and out of the city through ports and gates, and for places associated with travel by road and rivers (Rau and Zeller 2007).

4.4 Mapping Economical Activities

Mapping cities at the scale of quarters or parishes can be also improved by changing scale. Building scale mapping makes it possible to examine in detail how activities are distributed around specialized markets, for instance around slaughtering halls (skinners, butchers, candle makers...), weighing places, ports, colleges, justice halls, book making and selling (paper sellers, printers, writers...). Some specialized zones appear, where particular activities, sometimes in association are located.

4.5 Occupation Densities

The diversified use of urban space is linked to the greatly varying rent values, and corresponds to varying densities and incomes, from mono-familial houses in rural looking areas, then built in rows in the suburbs' main axes, to the multi-familial up to 6–7 stories buildings constituting the interior of the city and (at Lyons) where sometimes more than 100 people lived, blending huge apartments for the richest (400 m^2 and more) and mono-room ones of the poorest under the roof (less than 10 m^2). Censuses make it possible to map two fundamental variables, the number of people living in the same apartment or house, and the cohabitation coefficient (number of households living in the same dwelling unit). It is possible to study the neighborhood at its true scale, according to which measure the members of a formal sociability group, a professional one, or a confraternity, or the militia men were in the mean time close neighbors. It is also possible to scrutinize the informal social groupings, for example by examining the invitation lists (Zeller 2005). In a similar way, the study of estate patrimonies, from the tiniest to the largest ones, is of prime importance. The paradigm of socially determined behaviors is precisely mirrored by large speculative properties located according the opportunities of the market, patrimonies of artisans, traders and shopkeepers resulting partly from a homo-endogamy, concentrations of nearby buildings linked to a lineage, and the multi-ownership of buildings generally in the poorest areas; all being under the contingency of real estate values.

4.6 Socio-Spatial Relationships and Urban Geopolitics

Dwelling densities, neighboring relationship networks, proximity of the other: solidarities, and the conflicts likewise, are in part conditioned by the physical built-up environment. From this one, more curiosity emerges based on the police records: what is tolerated or forbidden in the uses of public space and shared spaces of multi-familial buildings (entrances, corridors, staircases) (Zeller 2005); what is known of the great riots and revolts, in a perspective of staseology contextualizing socially and spatially the transgressions. Obviously, people's reactivity, similar to that of the London mobs, is proportional to the dimensional structure of built-up and public spaces, and to traffic intensity in streets and squares.

Classifying each street according to the value of its bordering buildings and the nature of traffic helps us to understand the way institutional districts (quarters) were drawn. Generally, the quarters at Lyons were not made of assembled blocks, but mainly composed from street sections, connected sometimes to a small square where the militia could regroup. First of all to avoid the development of riots by implementing efficient supervision of the population, and then to repress them more effectively, the militia quarters were drawn in order not to isolate potentially dangerous popular areas. In the *longue durée* of the modern period, strategic modifications took place (Zeller 1990). The study of these changes is greatly facilitated by the use of GIS, as each building is associated to social and economic data on its proprietors and tenants.

4.7 Qualitative Aspects

A last possibility examined here is the possibility to confront written literature, travelers' narratives, reports, medical and administrative memoirs, planning decisions, in order to form a precise physical and quantitative knowledge of the city. Visible aspects of the architectural landscape and ambience of the street, smells, sounds, lights, may have conditioned these to be all the more interesting and informative as they are not isolated from reality.

5 Conclusion

Being able to locate at the building's scale, dwellings, activities, values, and space uses, opens immense new perspectives of research. Without precise plot and building plans, the social geography studies have remained somewhat crude, closed in a heuristic given, impossible to deconstruct. It was generally possible to process the data only at the parish or quarter scale, according to sources (Michel 1974; Pascal 1994; Junot 1997; Cabantous 1994, etc.). One can object that sometimes the administrative sections are small enough to constitute homogeneous social units, mainly when the spatial segregation was efficient and the dwelling mono-familial, but this is not very frequent, and the heterogeneity of the space use modes is obvious when one deals with the available sources. The social and economic signification of the quarters' areas is a postulate, a hypothesis that was methodologically disqualified (Cabantous 2008). The GIS allows determining the right study scale for a given question.

The spatialization at the building's scale, as it has been shown, actually opens up new research fields. It is certainly possible, in some source contexts, to attain to a spatialization at the dwelling or apartment scale, in three dimensions when considering the different floors. This is possible from the 19th century censuses, but the architectural documentation is very scarce, and allows generally only for some single building studies. This could lead to a solid study of vertical segregation and of gender repartition. The 1709 census strongly suggests a difference between a 'male city', at the level of the shops and workshops, and a 'female city' at the upper floors. Some 1677 buildings were clearly inhabited only by females, be it with a male at the ground floor, whose role can only be guessed, perhaps protection, or control.

The spatialization according to the buildings does not answer the needs of the open and public spaces studies that require other methods and a different cartography (Rau and Zeller 2007). But the GIS makes it possible to study the path of people in town through their successive apartments, and the evolution of the proprietors' residences, which tend to shift from inside their own building to the proximity, then to elsewhere according to an increasingly abstract mode of ownership, dictated by financial managing, and less by proximity with the tenants and the building considered as a familial patrimony. This is another aspect of a *longue durée* evolution becoming visible when comparing the situations for a quarter in 1677 and in 1853 for example (rue Buisson sector). The various moves of the silk workers are also quite interesting to follow, as the 1745 register records all the subsequent transfers of silk masters.

Historical studies have mainly dealt with parishes or quarters, but they are rarely concerned in detail the social space of a whole city (Concina 1984; Chauvard 1998, 2005). This is due to the heavy investment required by a *microstoria* extended to a whole city. It is true that this is somewhat paradoxical, but only the scale enlargement makes the change in the object mapped possible, according to a better correspondence between the registering scale in the sources and the buildings. The use of data along more than two centuries was necessary at Lyons because we cannot rely on better preserved sources (certainly a possibility in some other cities), but it makes nonetheless possible a better understanding of the trend scales, economic cycles of a few years, great historical context changes, *longue durée*.

A new paradigm emerges from the perspectives this use of the GIS opens. The spatiality of urban space cannot anymore be limited to 'a city' or 'a quarter', or 'a moment' determined by a given plan or a given source. The actual facts are too complex to be efficiently reduced to such generalizations or simplifications, and a way to overcome this generalization is possible. Urban space is not isomorphic, and has to be studied at the pertinent scale of the intervening events that shape it, nor is it fixed and knowable only at fixed dates, but it is transforming through time, which particularly the building permits show. This implies a multiplication of a factor 100 of the spatial units used for mapping, one quarter comprising, at Lyons, about 100 buildings, but the results are quite rewarding.

Nonetheless, mapping cannot be a scope *in se*. Making maps as a simple illustration of a discourse is naïve and obsolete, but mechanically transferring pre-built mass data on a map is no longer a good method, because no deconstruction is processed according to the right scale of phenomena and their right spatial nature. Text and map dialectics are to be placed at another level. The precise location of phenomena is essential information. It is the only way to contextualize according to land values, social occupations, and architectural structures. It is also a way to compile isolated information spread in many written sources, meaning nothing as long as it is not reunited and linked to precise place and time. Dynamic mapping allows a crossing of synchronic and diachronic data. Let's compare this to a textile frame, the weft threads being made of synchronic data, censuses, tax records giving a static state, the warp threads being diachronic data, architectural modifications, property mutations, changes of use. Diachronic cross-mapping clearly shows the shared logic of transformations. The real estate behaviors ("comportements immobiliers", Chauvard 2001), the transmission and transformation of the property values (Lepetit 1994), can only be studied efficiently through a precise spatial knowledge, only allowed for by the iteration between the map and the text.

References

Boudon F, Chastel A, Couzy H, Hamon F (1977) Système de l'architecture urbaine, le quartier des halles à Paris. Centre National de la Recherche Scientifique, Paris

Brac FPS (1787) Mémoire lu par M. Brac, avocat au parlement et aux cours de Lyon, ancien échevin de la même ville dans l'assemblée de Messieurs les anciens échevins de la ville de Lyon, tenue le mercredi 30 août 1786 dans l'hôtel des exercices de la Flèche, avec quelques notes ajoutées par l'éditeur. Manuscript, La Perrière private archive

Cabantous A (1994) Le quartier, espace vécu à l'espace moderne. Histoire, Economie et société 3:427–440

Cabantous A (2008) Histoire de la nuit. Publications de la Sorbonne, Paris

Cabestan J-F (1997) La naissance de l'immeuble d'appartements à Paris sous le règne de Louis XV. In: Rabreau D (ed) Paris, capitale des arts sous Louis XV. William Blake and co/Arts and Arts, Bordeaux, pp 167–195

Chauvard J-F (1998) Scale di osservazione e inserimento degli stranieri nello spazio veneziano tra XVII e XVIII secolo. In: Calabi D, Lanaro P (eds) La Città italiana e i luoghi degli stranieri. XIV-XVIII secolo, Laterza, Rome-Bari, pp 85–107

Chauvard J-F (2005) La circulation des biens à Venise, stratégies patrimoniales et marché immobilier (1600–1750). Ecole Française de Rome, Rome

Claval P, Claval F (1986) Cahors au milieu du XVIIe siècle et la géographie sociale de la ville pré-industrielle. In: Claval P (ed) Géographie historique des villes d'Europe occidentale, II—Espaces sociaux et paysages urbains. Publications du Département de Géographie de l'Université de Paris-Sorbonne n. 14, Paris, pp 69–72

Concina E (1984) Venezia nell'età moderna. Struttura e funzioni, Marsilio, Venice

Garden M (1970) Lyon et les Lyonnais au XVIIIe siècle. Les Belles Lettres, Paris

Gauthiez B (1995) Trasformazione dei valori urbani tra settecento e ottocento: il caso di Lione. Storia urbana 71:149–170

Gauthiez B (2004) Des unités pertinentes pour mesurer la ville concrète. *Histoire et mesure* XIX/3–4:295–316

Gauthiez B (2008) Lyon en 1824–32: un plan de la ville sous forme vecteur d'après le cadastre ancien. Géocarrefour 83(1):57–67

Gauthiez B, Zeller O (2006) Lyon aux XVII-XVIIIe siècles, la fabrique de la ville. In: Bajolet E, Mattei MF, Rennes JM (eds) Quatre ans de recherche urbaine 2001–2004, vol 1. Presses universitaires François-Rabelais/Maison des sciences de l'homme, Tours, pp 459–63

Gauthiez B, Zeller O (2009) Un SIG historique sur l'espace de Lyon aux XVII-XIXe siècles. In: Panzeri M, Faruggia A (eds) Fonti, metafonti e GIS per l'indagine della struttura storica del territorio. Celid, Turin, pp 39–49, 114–117

Gauthiez B, Zeller O (2010) Ordre textuel et ordre spatial à Lyon à l'époque moderne. Du parcours de visite au rôle nominal, une spatialité implicite. Histoire et Mesure XXV-1:197–230

Gauthiez B, Zeller O (2013) Le dédommagement des reculements: un instrument de la politique d'aménagement urbain à Lyon aux XVIIe et XVIIIe siècles. Histoire et Mesure XXVIII-1:45–73

Gregory IN (2003) A place in history: a guide to using GIS in historical research. Oxbow books, Oxford and Oakville

Gregory IN, Ell PS (2007) Historical GIS: techniques, methodologies and scholarship. University Press Cambridge, Cambridge

Junot Y (1997) Mixité sociale, habitat et propriété: la paroisse Saint-Jacques de Valenciennes en 1602 d'après un registre du centième. Revue du Nord LXXIXW: 413–427

Lepetit B (1994) The pre-industrial urban system: France 1740–1840. University Press Cambridge, Cambridge

Michel H (1974) Urbanisme et société à Montpellier sous l'Ancien régime. Un exemple: le sixain Saint-Firmin (1665–1756). Annales du Midi 116:21–43

Pascal C (1994) Bipolarisation sociale dans la ville d'Ancien Régime. Le sixain Sainte-Croix de Montpellier, 1665–1788. Revue d'Histoire Moderne et Contemporaine 41–43:395–417

Rau S, Zeller O (2007) Police des voyageurs et hospitalité urbaine à Lyon à la fin du XVIIe siècle. In: Burkardt A (ed) Commerce, voyage et expérience religieuse (Europe, XVIe-XVIIIe siècles). Presses universitaires de Rennes, Rennes

Schätti N, Viaccoz De Noyers A-M (2005) L'Atlas historique de la ville de Genève, vieilles méthodes, nouveaux outils. Patrimoine et Architecture 14–15:58–64

Zeller O (1990) Géographie des troubles et découpage urbain à Lyon (XVIème-XVIIIème siècles). In: 114ème congrès des sociétés savantes, Paris 1989. Les espaces révolutionnaires, editions du Comité des Travaux Historiques et Scientifiques, Paris, pp 43–59

Zeller O (1995) Enjeux d'urbanisme à Lyon en 1777. Propriétaires contre promoteurs. Bulletin du centre Pierre Léon d'histoire économique et sociale 1:3–15

Zeller O (2005) Espace privé, espace public et cohabitation. Lyon à l'époque moderne. In: Haumont B, Morel A (eds) La société des voisins, partager un habitat collectif, editions de la Maison des Sciences de l'Homme, Paris, pp 187–207

Mapping Long-Term Urban Space Structures: Barcelona as a Case Study

Manel Guàrdia Bassols and Sergi Garriga

Abstract The spatial dimension of demographic, economic and political processes is commonly disregarded or else it may happen that space is only considered when major urban projects or transformations of great morphological impact are involved. The systematic mapping of the data offered by available documents is much less common. Such mapping entails a laborious effort, but it is of great help in revealing the deep dynamics of a city that would otherwise remain concealed. Mapping is also useful to broaden one's capacity of observation and to pose new questions. Similarly, it provides a way of deducing how a method to ascertain how the people inhabiting the city in the past perceived their world. It should not be forgotten that our relation to space is fundamental to the construction of our individual and collective memory and, consequently, to the organization of our experience (Halbwachs 1952/1925; Pfeiffer and Foster 2013). The systematic application of manual mapping to the Barcelona of the 15th to the 19th centuries shows that, in the long term, space-society relations may be of explanatory value in the study of the form, the social and functional structure and, in short, the lived experience of the city.

1 Introduction

The spatial dimension is fundamental in many disciplines including archaeology, geography, sociology, anthropology and the history of architecture and of urban planning. Despite this, studies attributing genuine importance to space have been few, fragmentary and little known in general outside the strict limits of their respective academic fields. The scant use of mapping in urban history research is

M. Guàrdia Bassols (✉) · S. Garriga
Universitat Politècnica de Catalunya, Barcelona, Catalonia, Spain
e-mail: manel.guardia@upc.edu

S. Rau and E. Schönherr (eds.), *Mapping Spatial Relations, Their Perceptions and Dynamics*, Lecture Notes in Geoinformation and Cartography,
DOI: 10.1007/978-3-319-00993-3_6,

especially surprising; above all when it is considered that spatial relations are essential to urban life and therefore highly relevant for urban historians, and that their study is vital for a deeper understanding of cities. Systematic mapping may perhaps appear to be a clumsy and often excessively laborious approach, but it is probably the only way to bring the relations between space and society to the foreground, thus shedding light on social relations and dynamics which would otherwise go unnoticed. These urban dynamics highlight the various historical temporalities defined by Fernand Braudel and it is especially important to consider permanence, structural inertia and changes in long-term spatial relations.

2 Apparent Immobility of the Pre-industrial Urban Structure

The persistence of the city's plan and shape and the transmission of its morphology have been recurrent themes in the history of the city and the landscape. Many authors, including Marcel Poëte and Pierre Lavedan, have underscored the amazing permanence of European cities' street systems and even of their plot layouts. It is essential to go beyond the city's map, however, and to study the map's social contents. For this reason, in our research we have sought to relate the permanence and changes of Barcelona's morphological features to the permanence and changes of these features' social and functional contents. Tax documents such as hearth tax records from the 14th to the 16th centuries and the 18th-century cadastre tax enabled us to verify the street layout of the past on the basis of its mapping and then to go on to map the great wealth of demographic, social, economic and functional information that these documents provide on a long period of life in the city.

Only tax documents dating as from 1516 (beginning with the hearth tax for that year) provide enough information to allow the streets of the whole city to be plotted. Nonetheless, if the street layout from 1359 to 1716 (in the cadastre for that year) is taken into account, it can be seen there has hardly been any variation at all. Indeed, Barcelona is a classic case of permanence. In like manner, the tax documents for 1516, 1640 and 1716 allow the mapping and comparison of the hearths per hectare, showing that there has not only been a remarkable stability of the population's spatial distribution but also a close relation between the respective densities of the street layout and of hearths per hectare. The explanation almost certainly lies in the growth process undergone by Barcelona since the 10th century, which shaped the city physically, socially and functionally. It may be observed that, within the late medieval walls, the most populated areas were the part of the city located around the market outside the ancient Roman gate, the waterfront neighbourhood (where mariners and port workers lived), and the area near the Rec Comtal mills' channel (where wool and leather crafts were located). With its dense layout and population, this was the primary working area of Barcelona,

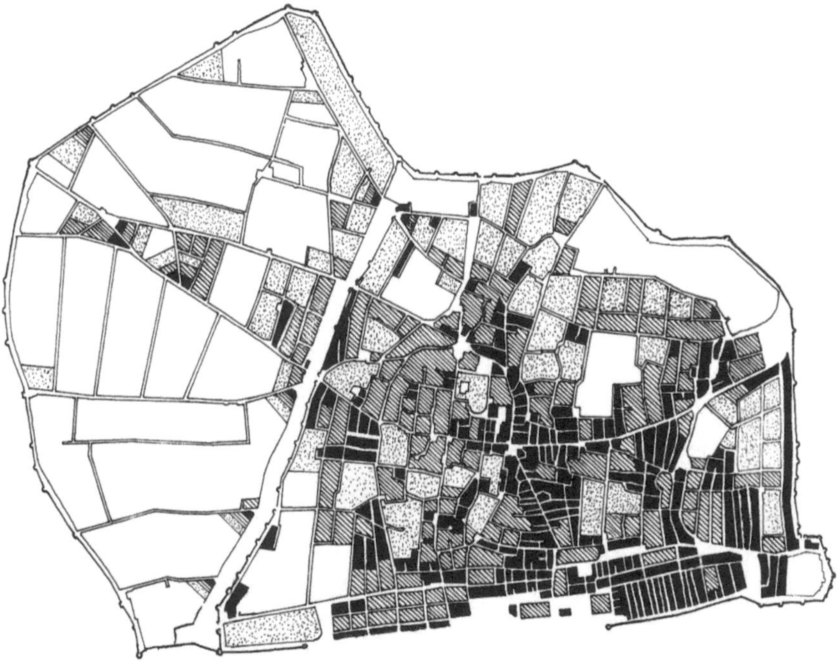

Fig. 1 Street layout density in 1516 (Hearth tax), <30 m between streets, 30–50 m between streets, 50–60 m between streets. >80 m between streets

accommodating 41 % of the inhabitants of the city in a space comprising only 14 % of its overall area. This shows that it can be quite useful to consider the street layout density as a basic reference for the morphological and historical study of medieval European towns (Figs. 1, 2).

The mapping of the tax documents' demographic, social, economic and functional information reveals the continuity and permanence of the demographic weights of the city and of its social and functional characteristics.

This stability or quasi-immobility of the pre-industrial city led to its study by means of models like those of Sjoberg (1960) and Vance (1971), which seek to identify the common aspects and critical variables of large sets of cities. Both postulated simple uniform models, describing the pre-industrial city as a walking place in which home and work were situated close together. They posited a city of simple uniform social districts; the central core was dominated socially by the homes of an élite class and additionally there was a number of occupationally distinct but socially mixed quarters as well as a residual population of very poor people living in the city's back alleys and on its fringes.

Gideon Sjoberg explained the preindustrial urban residential patterns by means of a rather simple uniform model that reflected the city's social pyramid, with a central core that was socially dominated by the élite class's homes clustered

Fig. 2 Hearth density per hectare in 1516 (Hearth tax) >150 hearths/ha 100–150 hearths/ha 50–100 hearths/ha <50 hearths/ha

around important religious, political and administrative institutions in the city centre. Around this core there were socially mixed quarters of craftsmen who were loosely clustered by occupations (with a poor population in these quarters' back alleys). The outer edge of the city, where there were vegetable gardens, was inhabited by farmers, a residual population of labouring classes and the very poor. Inhabitants were often grouped ethnically in various parts of the city within neighbourhood clusters.

James E. Vance Jr. considered that social differences in the medieval town were not horizontal because 'the residential structure was chiefly vertical'. Accordingly, he proposed a different city model formed by the aggregate of master craftsmen's houses with shops on the ground floor, the master craftsmen's family quarters on the first floor and, above, the rooms of the journeymen, apprentices and servants as well as the storerooms. This resulted in occupational zoning and social class mixing: the city was 'many-centred' with diverse craft quarters, each with its own shops, workplaces and each reflected a complete social spectrum. Families were located by occupational accident rather than by rent-paying ability.

Both models sought to draw attention to what they considered to be the critical variables and held that pre-industrial societies were divided into 'orders' or 'estates', which were legally rather than economically defined groups in sharp contrast to post-industrial societies. Further detailed historical research and

Fig. 3 Synoptic map of Barcelona, 1716 (Cadastre tax)

mapping of social indicators help us to redefine such generalizations and to pose new questions. For example, the research of Langton (1975) and Carter (1983) stressed the emergence of the retail centre as the backbone of the city's central area.

The study of Barcelona from the 15th to the 18th centuries shows that, from late medieval times, a specialized retail centre did indeed form the backbone of the city's central area. In fact there is a double centre reflecting the *Ancien Régime's* twofold social stratification based on the financial value of the land in the retail centre and the status value of centrality for the households of higher standing. After mapping the cadastral information from just after 1714, we developed a more accurate synopsis. Although there were some non-uniform or exclusive neighbourhoods, all the areas may be characterized by their predominant features. Briefly, a double centre and a periphery with differentiated parts may be distinguished, as shown in the synoptic social map of Barcelona based on the cadastre tax of 1716.

The black area was the city's retail backbone. Formed by Barcelona's most central and highly frequented thoroughfares, it was a dense area where the houses stood on small, but very costly plots and were inhabited by wealthy craftsmen, encompassing the most accessible and commercial parts of town.

Fig. 4 Master craftsmen's households. Barcelona, 1716 (Cadastre tax)

The hatched areas were low-density central parts of the city that were dominated by wealthy higher-status non-working households formed by rentier families with an income derived from property, a legacy or some other source, and who did not need to work. Intermingled within these areas were numerous poor families, often living in ground floor shops or in back alleys. In many cases they provided services for the wealthier families.

The dotted areas were densely populated and predominated by households of low economic standing. The neighbourhoods here were of greater density and were centres of industrial or port-related activities. Wool and leather craftsmen lived near the mills' channel whereas port workers and mariners lived near the waterfront.

The other peripheral areas (white areas) were of low density, with large neighbourhoods of low economic level including a substantial number of agricultural labourers and farmers (Fig. 3).

The 1716 cadastre tax makes it easy to identify and map the master craftsmen's households where the master craftsmen, their families, journeymen and apprentices lived under the same roof, as Vance's model suggests (Fig. 4). They are only found in the retail backbone of the city (the black areas on the synoptic map of Barcelona in 1714).

3 Urban Structural Inertia and the Great Impact of the Demolitions for the Citadel

In the period under study, many physical, social, economic and political changes took place. Perhaps the easiest of all these changes to explain are the demographic ones. It has generally been believed that in the pre-industrial city, the 'residential mobility within the town walls was quite limited, as there were few moves to locations beyond the walls, except in the very large towns, where suburbs had already developed' (Denecke 1988, 138). The study of Barcelona, however, shows a very high degree of mobility among the population, thus questioning this widely accepted assertion.

The Easter communion books of Barcelona's Sant Just parish recorded the number of communicants each year and allow an investigation of short-term population changes (Fig. 5). Specifically, the population of Sant Just parish fluctuated between 3,000 and 900 communicants (the latter figure during the Bubonic Plague of 1651). In this period, there was an average of about 2,100 communicants. This sudden population change highlights the fact that the pre-industrial city acted much like a sponge, absorbing people in growth phases and expelling them in phases of contraction. A study of the existing legal documents shows how buildings and dwellings were modified over time to accommodate a greater or lesser number of inhabitants. As the population grew, houses were divided to accommodate a larger number of families and when the city's population diminished, dwellings re-expanded by integrating rooms that had been abandoned.

The same parish communion books, which recorded the Easter communicants, shed light on the population's renewal by streets. We verified which surnames and forenames in the parish communion books of 1717 were also to be found in the books of 1693, 1700, 1707, 1715, 1719, 1724, 1729 and 1744 (Fig. 6). Except in very exceptional cases, the result was that only 40–60 % of the inhabitants remained longer than 2 years at the same streets. This was not only due to the high mortality rates but also to people's high level of domestic mobility, which often entailed a change of town to work as apprentices, journeymen, government employees or soldiers, or just to seek their fortune. The research of Bardet (1983) on the city of Rouen confirms this extremely high mobility, thereby challenging the very commonly held idea of the pre-industrial city's great residential stability. An analogy may be drawn between these demographic fluctuations and the phases of breathing: the effect of this 'demographic breathing' on the functioning and renewal of the regional urban system is worthy of study.

It should be stressed in any case that there is a sharp contrast between the numerous microscopic changes and the macro-structural stability. The 'demographic breathing' did not substantially affect the urban space's social and functional structure. Consequently, in the long term there were no outstanding changes on the whole in Barcelona's urban structure. From comparing the mapping of the tax documents of 1516, 1640 and 1716, it may be seen that the city remained

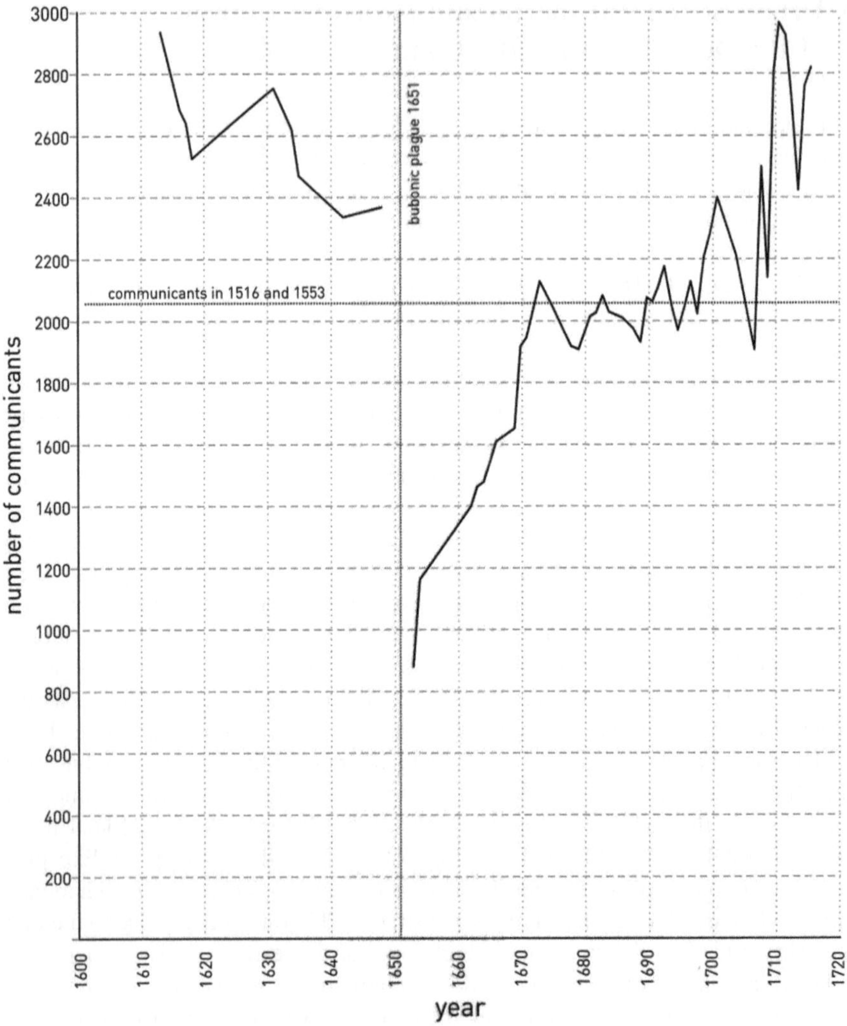

Fig. 5 Demographic fluctuations between 1610 and 1714 (Sant Just parish, Barcelona)

basically as it was. The demographic stability of the city's various sectors between 1516 and 1714 is indicative of the urban structure's great overall stability.

In this context, the large impact of a very important permanent change—the demolition carried out to make way for the Citadel—is of interest. This traumatic event allows one to test the extent to which inertia dominates pre-industrial urban structures. The building of the fortress and its esplanade in 1714–1718 after the fall of Barcelona at the end of the War of the Spanish Succession called for the demolition of a very large part of the city's most highly populated and active neighbourhood and the displacement of 17 % of the neighbourhood's population

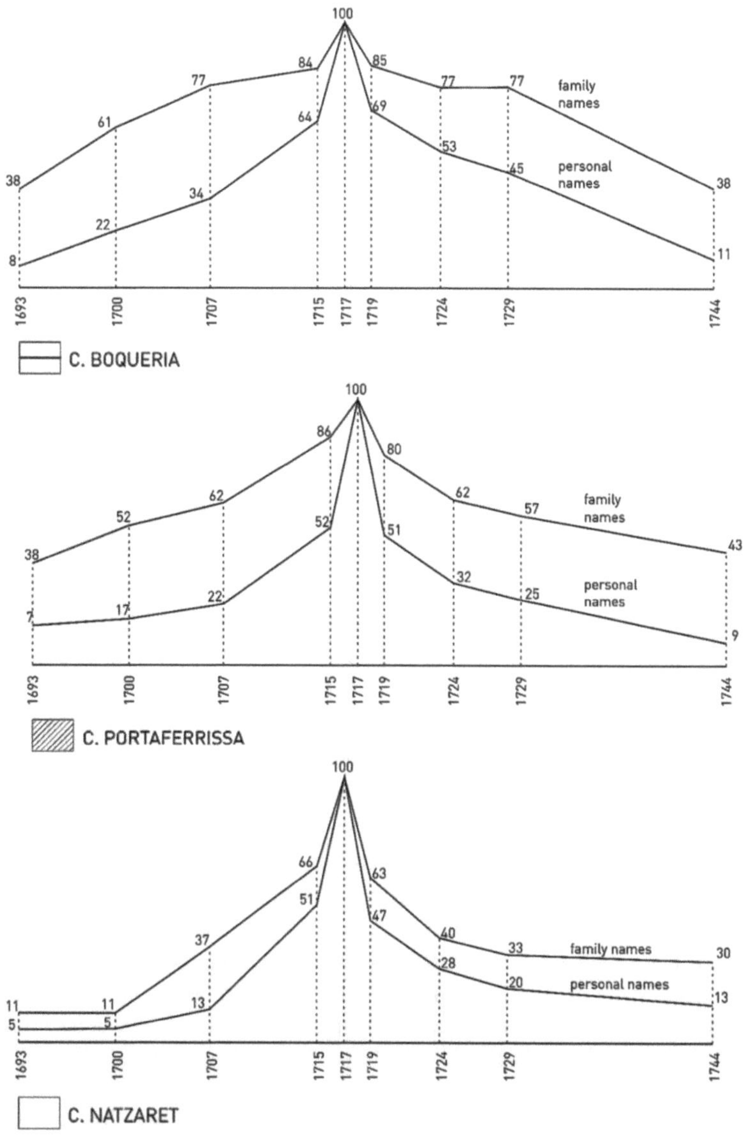

Fig. 6 Barcelona: Street mobility around 1715 (Cadastre tax 1716—easter communion books of Sant Just and Santa Maria del Pi parishes)

(about 6,400 inhabitants) (Fig. 7). The authorities intended to build two substitute neighbourhoods to accommodate the displaced families. One neighbourhood was to be on the waterfront for port workers, sailors, boatmen and fishermen while the other would be in the Raval (the city's historical suburb) for affluent displaced

Fig. 7 Demolition of a high-density area for the Citadel, 1716–1718

families. Both projects proved to be a complete failure, however. The displaced people did not trust the legal security of the proposed beach housing district, which would have had to be destroyed if the city was besieged; the Raval neighbourhood, for its part, was an unattractive location, situated far from the central area of urban activity. Only about 30 % of the displaced families can be tracked within the city and it has been found that they moved to places quite near the demolition area (Fig. 8). The rest of the displaced people probably moved outside the city walls where a shanty town arose near the port, or they took refuge in other towns. It was easier to move to another town than to a distant neighbourhood like the Raval.

Despite the great extension of the demolition area, the demographic growth from 1716 and the many changes in this period, the district densities of 1770 show that the population was concentrated in the same sectors as in the past, which highlights the resilience of the city's functional structure and the inertia in the ways of viewing and understanding this structure (Fig. 9). Consequently, the area of greatest density in the city remained close to the place where the demolition had been carried out. As from 1753, the newly planned Barceloneta district became the

Fig. 8 Families displaced by the Citadel demolitions in 1716

port workers' area, replacing the shanty town that had grown up outside the walls near the port after the demolition to make way for the Citadel.

4 Transformation of the Urban Structure Within the Medieval Walls

After two centuries of continuities and permanence, the building of the Citadel marked the onset of a long period of transformation. The true transformation of the urban structure inside the old walled precinct, however, took place above all between 1775 and 1850. In Barcelona, a sustained growth of the population, which was common to most cities, was accompanied by far-reaching economic changes that turned it into the Iberian Peninsula's leading industrial centre. The urban structure's transformation has often been focused on the gradual replacement of the so called 'craftsmen's houses' (comprising in one structure the workshop and residence of a single family) by 'tenement houses', which were taller (from five to seven floors) and contained apartments that were rented to a number of families. In

Fig. 9 District densities, 1770

addition to revealing the progressive separation between places of residence and work, it was considered that this transition reflected the mercantilisation of relations with a visible replacement of use value by exchange value. Other processes of equal or even greater significance, however, may be added to these gradual changes. The first of these processes involves the progressive displacement of the city's population and activity towards the Rambla and the Raval, an area that had been traditionally considered peripheral. Although the Raval stood within the walled precinct, it was very sparsely inhabited because it stood outside the first medieval wall and had a clearly suburban character. The attempts to carry out its urban development shortly after 1714 failed and, as may be seen from the density of the districts in 1770 (Fig. 9), the city's activity continued to unfold almost exclusively in the neighbourhoods nearest the Citadel. As from this year, however, a long-term change in this trend began to emerge. The critical event in this turning point, which marked an end to the century-old inertia, was the Rambla's urban development, which began in 1775. This process entailed the demolition of the city's first medieval wall and the building of a new tree-lined promenade that changed the standing of the whole Raval district. From 1785 to 1859 many new streets were built and inhabited there and the city's growing densities brought about a complete change in its population's spatial distribution (Figs. 10, 11).

The new cotton mills established in Barcelona from the 1740s were initially located in the traditionally productive areas of the medieval city, but by the end of

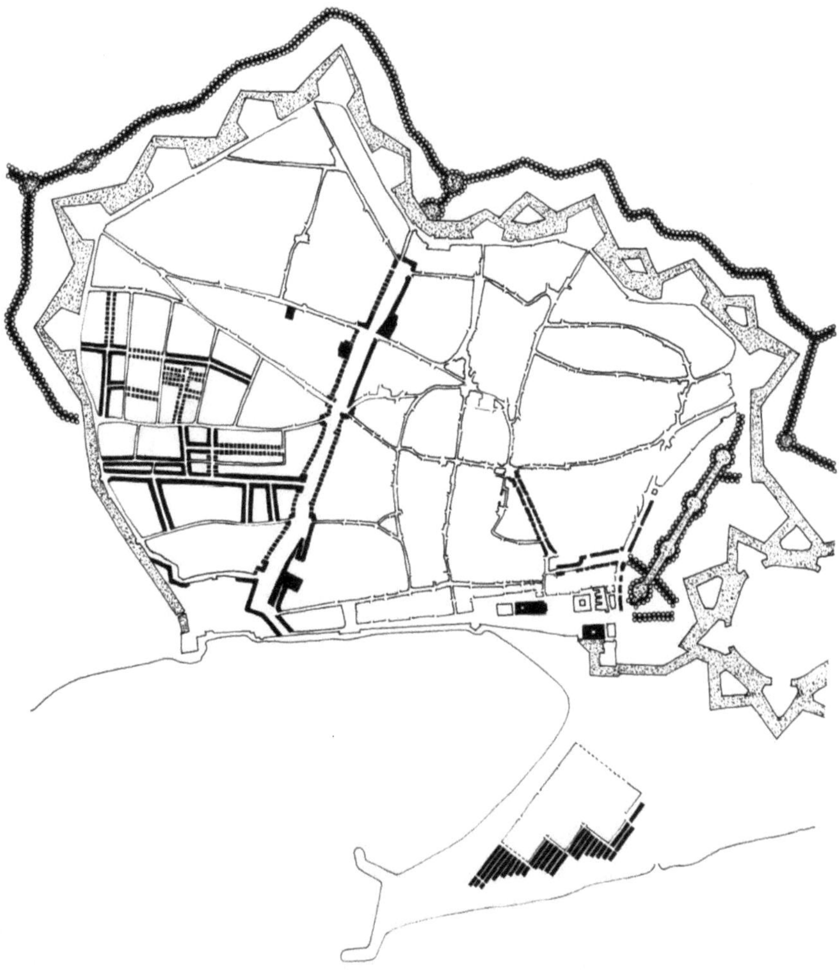

Fig. 10 New streets built in 1785–1808

Fig. 11 New streets built in 1836–1859

the 18th century most of these mills and above all the large ones with the greatest need of space had installed themselves in the Raval district. When steam engines were introduced in 1833 and the industrial impetus was renewed after the crisis of the early decades of the 19th century, the Raval's industrial character became even more pronounced. This is shown by the location of the various activities of the cotton industry according to the data of a guidebook from 1841 (Fig. 12).

The introduction of steam-driven industry coincided with political changes that put an end to the *Ancien Régime's* foremost structures and completed the city's transformation process within its old walls. Indeed, as from 1836, laws decreeing

Fig. 12 Cotton industry, 1841

the confiscation of Church properties allowed the replacement of religious buildings by new public spaces and services (squares, markets, university, theatres, new parish churches, jails, etc.). A change in the general perception as to what constitutes the notion of "city" had been taking place throughout Europe from the end of the 18th century: there had been a shift from a 'heraldic' view that valued cities for their antiquity, nobility, privileges and pious manifestations to a new view of 'service cities', a view that valued their economy, administration and public places and facilities (Teyssot 1977; Lepetit 1979; Monclús 1989). The change in the way of looking at the city laid the ground for the interventions carried out between the confiscation of the monasteries in 1836 and the demolition of the city walls in 1854.

The data compiled in *Teoría general de la urbanización* (General Theory of Urban Development) by Ildefonso Cerdà, the author of Barcelona's expansion project, provides a useful description of the city in 1859 (Cerdà 1867). Higher densities were concentrated in the peripheral areas where industries were

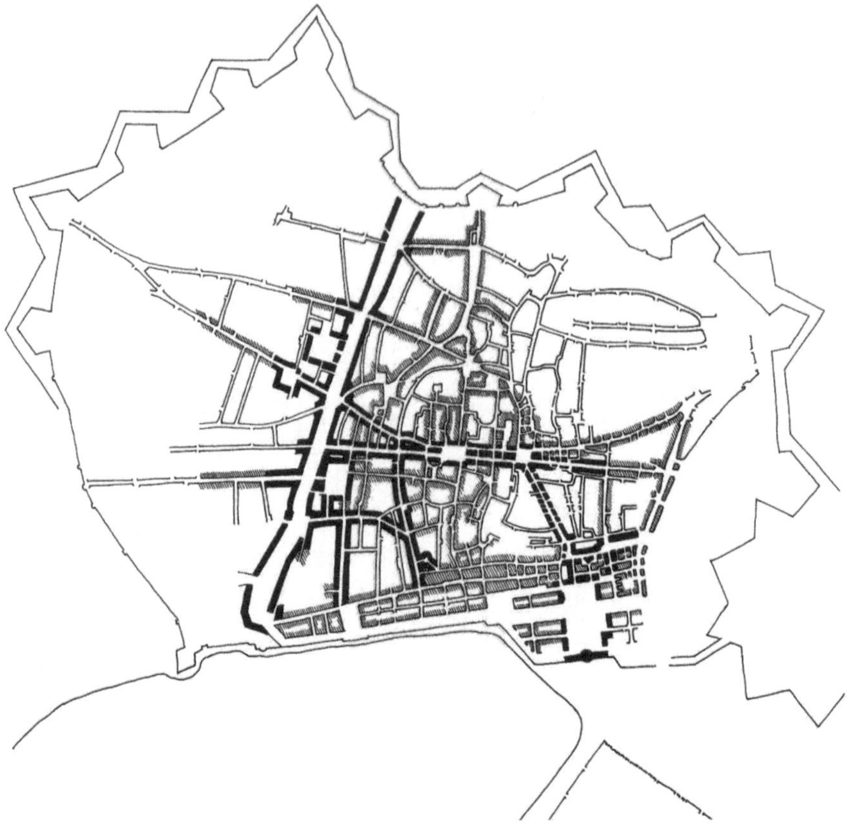

Fig. 13 Street land values, 1859

established and industrial worker families settled. The mapping of the cotton industry activities in 1841 contrasts with the higher street land values in 1859 (Figs. 12, 13). The political and economic changes between 1836 and 1859 led to a centre/periphery organisation of the city that was based mainly on land prices and rent-paying capacity, reflecting a purely economic stratification.

The struggle for the demolition of the city walls and the spectacular expansion of the city beyond its walled precinct as from 1859 often cause the primordial transformations of these preceding phases in Barcelona to be overlooked.

References

Sources

Hearth tax 1516 (Arxiu Històric de la Ciutat de Barcelona)
Cadastre tax 1716 (Arxiu Històric de la Ciutat de Barcelona)
Easter communion books of Sant Just parish (Arxiu Diocesà de Barcelona)
Easter communion books of Santa Maria del Pi parish (Arxiu Diocesà de Barcelona)
Francisco Zamora. Noticias para la historia de Cataluña. Manuscript 2.469 (Biblioteca del Palacio Real, Madrid)

Bibliography

Bardet JP (1983) Rouen aux XVIIe et XVIIIe siècles: les mutations d'un espace social. Societé d'Edition d'Enseignement Supérieur, Paris
Carter H (1983) An introduction to urban historical geography. Edward Arnold, London, pp 150–162
Cerdà I (1867) Teoría general de la urbanización. Tomo II, Imprenta Española, Madrid
Denecke D (1988) Social status and place of residence in preindustrial german towns: recent studies in social topography. In: Denecke D, Shaw G (eds) Urban historical geography: recent progress in Britain and Germany. Cambridge University Press, Cambridge, pp 125–140
Garcia-Espuche A, Guàrdia M (1985) Espai i societat a la Barcelona preindustrial. La Magrana, Barcelona
Garcia-Espuche A, Guàrdia M (1986) Un intervento urbano nel 'tempo lungo': la Citadella di Barcellona, 1715–1869. Storia Urbana. 37. Franco Angeli, Milano, pp 3–28
Garcia-Espuche A, Guàrdia M (1992–1993–1995) Chapters on urban structure in volumes 3-4-5-6. In: Sobrequés J (ed) Història de Barcelona. Enciclopèdia Catalana, Barcelona
Guàrdia M, Monclús FJ, Oyón JL (eds) (1994) Península Ibérica. Atlas Histórico de las Ciudades Europeas. Salvat, Barcelona
Halbwachs M (1952\1925) Les cadres sociaux de la mémoire. Paris, Presses Universitaires de France. Les Travaux de l'Année Sociologique, Paris, F. Alcan
Langton J (1975) Residential patterns in pre-industrial cities: some case studies from seventeenth-century Britain. Trans Instit Brit Geogr 65:1–27
Lepetit B (1979) L'évolution de la notion de ville d'après les tableaux et les descriptions géographiques de la France (1650–1850). Urbi II, Paris, pp XCIX–XVII
Monclús FJ (1989) Teorías arquitectónicas y discurso urbanístico. De las operaciones de 'embellecimiento' a la reforma global de la ciudad en el s. XVIII, vol 79, pp 25–40. Ciudad y territorio, Madrid
Pfeiffer BE, Foster DJ (2013) Hippocampal place-cell sequences depict future paths to remembered goals. Nature. doi:10.1038/nature12112
Pinol JL (ed) (1996) Francia. Atlas Histórico de las Ciudades Europeas. Salvat, Barcelona
Sanpere-Miquel S (1911) Los terrenos de la Ciudadela, Libro en donde queda sentado el valor que han tenido las casas que se han derribado para la Real Ciudadela y su Esplanada... Prueba 3.1911. Heinrich, Barcelona
Sjoberg G (1960) The pre-industrial city. Free Press of Glencoe, Glencoe (Illinois)
Teyssot G (1977) Città-servizi. La produzione dei *bâtiments civils* in Francia (1795–1848), vol 424, pp 56–65. Casabella, Milan
Vance JE Jr (1971) Land assignment in the precapitalist, capitalist and postcapitalist city. Econ Geogr 47(2):101–120

Part III
Mapping Humanities

Writing Spatial Relations and Dynamics: Movements in Urban Space (Barcelona, 16th–19th Century)

Susanne Rau

Abstract In this contribution, I intend firstly to show at a theoretical level and then demonstrate in a practical way the advantages of viewing urban space from the perspective of the historical actors. Here we are more concerned with applying urban space dynamically, which means we no longer view the perceived and described spaces as something given nor do we set them in rigid maps. Moreover, our view is focused, firstly more upon the perception of movement in urban spaces, or changes in this area, and secondly, on the awareness of hierarchies in urban spaces (in dichotomies such as centre–periphery, old–new, inner–outer) and, if appropriate, hierarchical changes during the course of time. Finally, in this context, the following question arises: is it at all possible to map movements within urban space or to map the perceptions of urban spatial transformations including urban hierarchies together with any concomitant changes?

1 Introduction

Although the following contribution is once again dedicated to the city of Barcelona,[1] the focus has shifted. It is now less a case of concentrating on the structures of urban space or the historical development of urban space as recently portrayed in a very succinct way by Casellas (2009), but instead we are now focusing more on the perception, use and description of urban space from the perspective of the historical actors themselves. However, this approach could be described as subjective, and thus be rejected as a non-scientific method, but this would be a gross misconception

[1] Cf. the essays in this volume by Manel Guàrdia Bassols (Mapping Long-Term Urban Space Structures: Barcelona as a Case Study) and Ekkehard Schönherr (Elements of Historical Knowledge About Urban Spaces. Reflections on the Requirements for a Dynamic Map).

S. Rau (✉)
Geschichte und Kulturen der Räume in der Neuzeit, University of Erfurt, Erfurt, Germany
e-mail: susanne.rau@uni-erfurt.de

S. Rau and E. Schönherr (eds.), *Mapping Spatial Relations, Their Perceptions and Dynamics*, Lecture Notes in Geoinformation and Cartography,
DOI: 10.1007/978-3-319-00993-3_7,

because what is being dealt with here is how we can study the subjectivity of the actors from a purely scientific perspective. In this contribution, I intend firstly to show at a theoretical level and then demonstrate in a practical way the advantages of viewing urban space from this perspective. Here we are more concerned with applying urban space dynamically, which means we no longer view the perceived and described spaces as something given nor do we set them in rigid maps. Moreover, our view is focused, firstly more upon the perception of movement in urban spaces, or changes in this area, and secondly, on the awareness of hierarchies in urban spaces (in dichotomies such as centre–periphery, old–new, inner–outer) and, if appropriate, hierarchical changes during the course of time. Finally, in this context, the following question arises: is it at all possible to map movements within urban space or to map the perceptions of urban spatial transformations including urban hierarchies together with any concomitant changes?

2 Urban Structure—Perception—Description

In what follows, my starting point is that the best way to perceive towns is to see them as spatial configurations, which—both concretely and figuratively—have been constructed as well as being perceived, experienced and represented (Rau 2013: 153–155). This approach differs from a purely historical architectural perspective in so far as this method is more concerned with the use made of urban spaces and their significance as well as with the formation of spatial relations and hierarchies arising from social interaction. Thus it distinguishes itself from any perspective based on the constitutional history of the city, mainly concerned with questions of autonomy and civil law. At the same time, this perspective is not be confused with the very recent, innovative media-based historical approach because our interest is not confined to the medial constructions or to representations of towns and cities, but we are more interested in the perceptions and practices of individual actors or groups as well as noting any urban spatio-temporal changes. This actor-centered viewpoint is fully aware of the fragmentary nature of human perceptions as well as any possible contradictions in behavior while at the same time it makes no attempt to reconcile these contradictions.

Since the advent of digital methods to present information visually and to map historical spatial relationships, a technique which is by no means a closed door to historians (cf. Knowles 2008), it would seem to be obvious that a geo-information system had be set up for this purpose. Usually, a huge research input is necessary in this medium to be able to do justice to the presentation of the comings and goings, observations, interpretations, creation and destruction of the towns and cities by the historical actors.[2] This is why in the first instance I chose to stick to a

[2] Nevertheless, at the University of Erfurt, we have digitalized individual travelogues in a three-year project funded by the Deutsche Forschungsgemeinschaft (DFG). The digital maps comprise

more traditional, text-based method and to rely on the insights of Henri Lefebvre (1901–1991), the philosopher and urban theoretician, and also on Michel de Certeau S. J. (1925–1986), a philosopher of culture.

At a very fundamental level, Lefebvre contributed to the introduction of the category of space into Marxian social analytical theory. In the meantime he has become widely recognized in the field of the sociology of space. It was important to him to stress the fact that space is not a pre-set, given entity but something that is socially constructed. For his analysis of social spatiality, he had proposed regarding space at three levels, which can intersect, support or contradict one another: *espace perçu* as space as it is generally perceived, felt and used, *espace conçu* as the envisaged space which had been deliberately conceived and planned and *espace vécu* as the space one actually experiences (Lefebvre 2000: 48–49). De Certeau's main interest focused on the daily praxis of the actors in their urban space (de Certeau 1990: 170–191, cf. also Füssel 2013). His higher aim was to show that individual actors do indeed possess power and are not merely agents subject to a higher disciplinary power, as it has been *bis dato* described by Michel Foucault. De Certeau proposed making a distinction between place and space as a basis for a set of analytical tools: only by the activities and dealings of the actors would (inanimate) places be transformed into spaces. This concept was neatly formulated in his now famous saying: "l'espace est un lieu pratiqué" (space is a place which is experienced) (de Certeau 1990: 173). He made a distinction between map (*carte*) and tours (*parcours*) to be seen as two basic approaches to space. The map represents a more panoptic, topographical presentation of space whereas the "parcours" (tour) envisages space—in a more concrete and descriptive way—by movement to and fro with the actors' comings and goings.[3] These alternative conceptions of urban spaces are not intended in any way to replace standard approaches based on town planning, or constitutional or social history, but instead, to complement them. The two theoreticians, however, were of the opinion that a town or city could not be fully represented from the perspective of a space planner, cartographer or urban sociologist.

Why, of all places, was Barcelona selected as a field of investigation? For the majority of people, football is probably the first idea that comes to mind in association with this city. It is also well known that this Catalonian city on the Mediterranean has during the last few years become a top-ranking tourist destination: in 2011, 7.3 million tourists visited the city and 15.5 million overnight stays had been listed for the same year.[4] This increasing interest is probably

(Footnote 2 continued)
the first stages for producing a digital atlas of towns and cities. Cf. Geschichte and Kulturen der Räume in der Neuzeit, URL: https://www.uni-erfurt.de/index.php?id=23985. Accessed 6 Aug 2013.

[3] For an application of this concept to the mapping of two early modern urban spaces (Geneva and Lyon), see Rau (2011).

[4] Tourists and overnight stays, 2007–2011, URL: http://www.bcn.cat/estadistica/angles/dades/anuari/cap13/C1302010.htm. Accessed 4 Aug 2013.

connected with the successful marketing of Barcelona as a tourist destination since the Olympic Games of 1992. A major factor for the choice of Barcelona is the fact that spatial layers and material from a variety of epochs have been preserved in the city, ranging from one century before the Christian Era up to the present time, even though they were appreciated with varying degrees of interest by the travelers. What, however, is somewhat less generally well known is the fact that Ildefons Cerdà's (1815–1876) city plan for the extension for Barcelona (1859) can fairly be regarded as the moment of birth for the field of modern town planning and that, in the wake of this project, Cerdà developed his *Theory of Town Planning* as well as a *General Theory of Urbanization*. According to the plans, the area of Barcelona was to be extended on a large scale and to be open to constant expansion; *de facto* only a small part of the plan was realized during Cerdà's life-time. The history of this planned extension can be regarded as well researched even though there is still a need to integrate a perspective on the history of perceptions of the city (Rodríguez-Lores 1980; Sabaté Bel 1985; Torres i Capell 1985; Zimmermann 2000: 149–165; Santa-Maria Batlló 2009). In 2009/2010 on the 150th anniversary of Cerdà's City Plan, Barcelona celebrated the "Any Cerdà" (Cerdà Year) with exhibitions, guided tours and conferences. This is a sign that the history of the Great Plan together with its concomitant alterations of urban forms has now even been incorporated into the city's marketing strategies and that the history of the city's transformation is now being marketed as an element of its identity: the updating of the historical city as spatial constellations in motion to the present time. This assumption can by no means be taken for granted with regard to towns whose transformations would be an interesting topic for research.

A brief glance at a modern city plan will suffice to recognize the present structure of the city. Casellas divides the city into roughly three areas: in the first instance, the old medieval city centre is still intact with its numerous old buildings and its remaining narrow, winding streets (Ciutat Vella). The Eixample forms the second area, which is a mainly residential quarter and arose as part of Cerdà's extension plans. The third area comprises firstly, the former villages which were being incorporated into the city from the middle of the 19th century onwards and secondly, the new city quarters which started being built in the 1960s. Also the latest alterations in the marina area can hardly be missed (Casellas 2009: 816). This structure was not always as it is now (Reimann 1996). In what follows, it will be shown what impression the city made on the eyes of the beholders who observed the city before its great extension.

It goes without saying that suitable sources are a prerequisite for investigating observations made during earlier periods of time on spatial relations and their changes: the first requirement of these sources is that they should deal directly with the topic of spatial relations in urban societies; the second is that such sources should be spread over a long period of time. Up to the present time, descriptions of urban space can be found in municipal chronicles, historical topographies or chapters taken from descriptions of whole countries. Usually these are official documents, often commissioned by the relevant authorities or frequently written as encomia to the said authorities in order to gain some office or position or even for a regular payment (a kind of grant). A completely different kind of source material comprises the so-called ego-documents such as letters or diary entries containing opinions on building projects, written by the town's inhabitants. However, it cannot be relied upon that either historical topographies or ego-documents will be available in sufficient quantities over a suitably long period of time. The sources which best fulfill these conditions are travelogues. Even though they have the disadvantage that they are not usually written by the city's inhabitants, but instead by strangers, it can generally be seen to be the case that the local people have a much more limited view of urban spaces than strangers to the city who have to rapidly find their bearings on arrival at unknown territory. This necessity generates a completely different awareness of the spatial relationships within the city. Secondly, these sources contain data supplied by the local people, which entered into the travelogues via a variety of communication channels—ranging from conversational exchanges to reading material such as almanacs or chronicles. In addition, this data would also go the rounds via a great variety of channels to diverse destinations including their readers, other travelogues and various books. In the 18th century, particularly during its second half, the number of titles had increased to three to four times the number compared with the previous century so that this could be regarded as a mass literary phenomenon (Roche 2006). In addition, there was the secondary use and circulation of this literature. Thus, at least in the 18th century, a discourse had taken shape, in which numerous male writers and a few female authors refer to, copy from or correct one another. The descriptions of the city contained in these travelogues should not be used to reconstruct the physical structures of the city at various points in time as this source is not primarily suitable for this task. Instead, the travelogues should, following Michel de Certeau, be seen as narratives of space (de Certeau 1990: 170–191; also Rau 2011: 173, 2013: 113–114, 178), thus as texts, subject to the narrative form, which, on account of their distribution, are capable of influencing the discourse on the city, also on any spatial relations and various changes taking place.

For our research purposes roughly 450 travelogues from various sources and languages (German, Italian, French, English, Spanish) covering a period of time from the 16th to 19th centuries were used with the great majority issuing from the 18th and 19th centuries. The constitution of this corpus has been described in another place.[5] Suffice it to state at this point that the comprehensive coverage of this bibliography is utopian and will continue to be so. The bibliography is, however, sufficiently dense in material so that both repetitions and deviations in the documents can be established. The reports containing usable descriptions of Barcelona's city space were then reduced to 150. And even here the majority has been taken from the 18th and 19th centuries. In the assessment stage, special notice was taken of the descriptions of the whole city, and the following significant alterations were noted: the building of the Citadel as a result of Barcelona's defeat

[5] Bibliography for the city of Barcelona, URL: https://www.uni-erfurt.de/projekt-geschichte-kulturen-neuzeit/bibliographien/. Accessed 6 Aug 2013.

at the end of the Spanish War of Succession (1701–1714), the construction of Barceloneta (1753)—the area situated on the city's built-up peninsula where the workers at the harbor and the military lived—the transformation of the Rambla into a promenade, the development of the Raval, an area in the western part of the city which for a long time had remained a relatively uninhabited quarter, the inner city conversions (sites made into esplanades and places), the demolition of the Citadel, the Cerdà project and finally, the projects which were not carried through.

3 Observations on Space as Described in the Travelogues of the 16th and 17th Centuries

This was connected with our first result—and at the same time it put into perspective and gave concrete form to Michel de Certeau's rather generalized thesis: the travelers of the 16th and 17th centuries paid scant attention to spatial elements or relations of the city, but certainly, the urban topographies in the travelogues on hand did not exactly play a major role. There was some reference to spatial relations in the reports, but their main focus was on data concerning distances between the towns and other locations on their journeys (in terms of time or miles), in other words, on their itineraries, and, at times, on their arrival at the city gate and, finally, on specific locations within the city.

In the memoirs of Sigmund Freiherr von Herberstein (1486–1566), who passed through Barcelona on a consular mission in 1519, hardly any reference to the city is made, but he was more concerned with the diplomatic negotiations with the then King Charles, the future Holy Roman Emperor. The only city places mentioned was the harbor and an inn (with no further reference), although there is more information on the place of disembarkation because Barcelona did not have a harbor, but just "ain Sannd unnd seicht" (sandbank and shallow water) (Chmel 1846: 39), so that the ship could not land there and so the travelers had to be carried ashore on the backs of crew members (Fig. 1).

In the middle of the 17th century, the Württemberg Privy Counselor Hieronymus Welsch (1612–1665) described his entry into the harbor when the party was welcomed with volleys. After giving a brief description of Catalonia's location, the writer then mentions its capital city by referring to pleasure gardens, the Academy, towers, the harbor, the Palace of the Viceroy, armory, Lonja (exchange) and weighing house, without, however, giving any information on their geographical location nor on how the buildings stood in relation to one another. Only the harbor (on account of the ships and merchants) and the Palace (in which the travelers were allowed to stay) gave the impression of being 'lively places' in de Certeau's use of the term (Welsch 1658: 230–231). There is a host of other such examples to support the impression that the travelers of the first half of the Early Modern Period made little effort to integrate urban topography or defunct thoroughfares into their narratives. This impression, however, contradicts the present

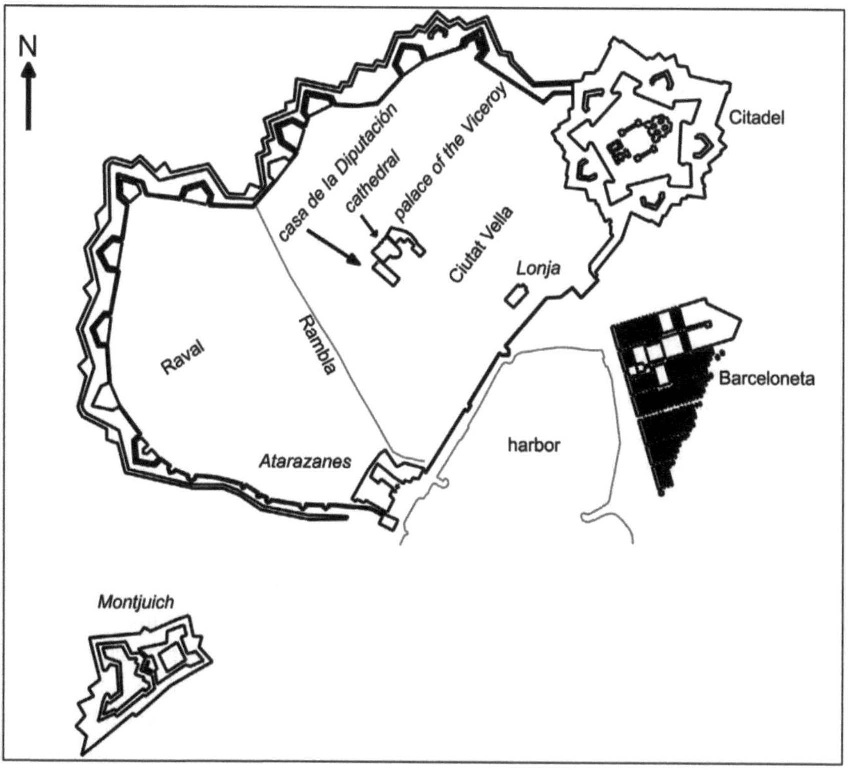

Fig. 1 Places and areas visited by early modern travelers—map: Leif Scheuermann, 2013

state of research which—as far as the towns and cities are concerned, both in general (Lestringant 2012: 8) and in particular (Conley 1997)—reflects the fact that the Renaissance claimed to prioritize space over time. In my opinion, this applies particularly to the geographical writers, to global travelers as well as to the makers of global maps (which had to be adequate for navigational purposes), even though this does not apply to all types of travel writers.

It is necessary to look around for other writers to learn more about urban spatial elements. On the one hand, geographical writers provide a useful source of information and on the other hand, the same applies to travelers who were not involved in diplomatic missions. An example of this type of writer of travelogues was Hieronymus Münzer (about 1437–1508), a physician and geographer, who, at the end of the 15th century, collaborated with Hartmann Schedel in Nuremberg to produce his world chronicle. From 1494/1495 he traveled to the Iberian Peninsular. Setting off from Gerona, he reached Barcelona and immediately described the city's location as would only seem to be appropriate for a Humanist: situated by the sea, framed in by an 'amphitheater' formed by the mountains; the city itself is surrounded by a fortified wall in the middle of which arises a cathedral. On the outside of the wall, but within its vicinity there are about 30 religious institutes (Münzer 1991: 7–21). On the inside of the walls he pays particular attention to the Lonja, the Casa del Infante (the Infante's Residence) and two Menorite monasteries. He was invited to the homes of two German merchants who were temporarily based in Barcelona. At the end of the travelogue, the city's subterranean, invisible space, i.e. the sewage system (*las alcantarillas*) is mentioned. As a whole document, the Münzer text can be seen as a panoptic cartographic description of space. The city and its environs are described in terms of concentric circles (wall—monastery belt—mountains) with the Cathedral at the centre, thus symbolically raising this building to the religious centre of the city. Individual city locations (the Lonja or the monasteries) are only slightly lively (on account of anonymous merchants, monks or nuns). Only the houses of the German merchants appear as lively places, in which Münzer and his fellow travelers were invited to a banquet and to a musical evening where there was a lot of eating, drinking, listening to music, dancing and making friends. However, nothing is written about the location and appearance of these houses.

In contrast how much more informative with regard to urban spatialities is the travelogue written by Thomas Platter (1574–1628), a student from Basel, who at the end of the 16th century had travelled through France, Spain, England and the Netherlands. Platter's travelogue begins with a review of the city's history since its foundation and then goes on to describe both the state of Catalonia and the city's location (Keiser 1968: 337–354).[6] Platter likens the city's shape to a half moon and seems to be delighted with the 2,000-m-long city wall, along which runs a pleasant footpath ideal for walks.[7] After that, the reader is guided through the city as if following a parcours: starting from the harbor and Montjuich he goes across the numerous squares and on to the "kaufhauß" (emporium) where he had to pay customs duty, then continuing through city streets, he pays particular attention firstly, to a tailor, then to a surgeon and finally, he proceeds onto the "Casa de la Deputacion [= Diputación]". In an another extract, he visits social venues such as houses of ill repute, taverns, wine drinking places, inns, and then onto the open air comedy theaters for which special "plätz (*theatra*)" (places) were set up (Keiser 1968: 347), and finally, onto the hospital. In the "neüwen oder vorstatt" (new quarter or suburb) (Keiser 1968: 348) he watched a French tightrope walker for some time. On his return to the historical city centre, he visited churches, then the medical college, which particularly interested him for obvious reasons (he had

[6] Platter based his historical details on the "Städtebuch" by Georg Braun and Franz Hogenberg (Keiser 1968: 338).

[7] "Es ist diese statt wie ein halber mon gegen dem meer gebauwen, welches an die stattmauren schlachet gegen levant, unndt ist ein überauß schöner spatzier weg in der statt an der mauren gegen dem *lustigen* meer *gestadt* hinaus erhöhet, daß man auf der stattmauren sitzen kan; ist bey fünfzehn hundert schritt lang unndt ettwan 20 schritt breit, darauf auch, ohngefahr in der mitte, des (viceroy) königs statthalters behausung mitt einer galerien biß an daß meer hinauß gebauwen ist, da man auch albereit hatt angefangen, ein ander port zemachen, wie folgen wirdt." (Keiser 1968: 339)

gained a medical degree in Montpellier). Platter is, however, not just a passionate walker, but he also had an eye for city structures and for changes in urban space. He regards the city he described as half-moon shaped and divided into two parts: "Unndt ist diese statt in zwo stett abgetheilet, die alte unndt die neüwe oder die vorstätt."(And this city consists of two parts, the old part and the new or the outskirts) (Keiser 1968: 342) By the 'new part' he was probably referring to the Raval which at that time was only thinly populated. Platter also had an eye for extensions of the city which had taken place in the past and he was able to discern past periods from architectural remnants: he wrote that it could be seen from the old town walls that Barcelona used to be a small town, which had twice been extended since that time and that the city had had thick walls and splendid towers twice built around it.[8] A few lines later, strolling between the fertile gardens and the pleasure houses, he makes the following point: if the old town wall were to be removed, you would never think that there had ever been two towns (Keiser 1968: 354). In this way, he may well have maintained that the former division did not play a role in the everyday lives of the town dwellers. One final point can be noted with regard to Platter's narration on space because, with regard to the harbor, an information gap is revealed between Herberstein's reports on the one hand and Welsch's travelogue on the other: he points out that the harbor had been relatively recently extended.[9] Foreign galleons and large ships could now land here, which was advantageous for the city's trade.

Platter's travelogue is the first one to deal with movement in urban space in a double sense, firstly, with regard to alterations in the city's architectural structures and secondly, with reference to bodily movement through urban space. Both these are narrative constructions; the first, because the writer makes his own selection amongst the transformations (he describes forms and often gives his own assessment of them); the second because the narrated parcours hardly corresponds to the one he actually went on, as Platter was in Barcelona altogether for only six days, and so must have viewed the city in stages whilst at the same time creating the impression that he had completed one single (uninterrupted) walking tour. There are other examples of these movement descriptions in 18th and 19th century travelogues.

---

[8] "Als wier zu Barselona wahren, gungen wier für die alte statt hinauß, *dann man noch an den alten mauren gespüret, daß sie vorzeiten ein klein stettlin gewesen unndt vom gestad des meers eins roßlaufs wegs abgelegen; hatt vier porten gehept mit hohen mauren, welche einer gleichen weite von einander stehen undt sich zu der welt vier örteren richteten, welche yetz aldo mitten in der statt stehen. Da aber die einwohner an der menge aufwuchsen, ist die statt zweymahlen mit zierlichen heüsern unndt herrlichen kirchen vermehret, ist auch zweymahl mit gewaltigen mauren unndt prächtigen thürnen umbzogen.*" (Keiser 1968: 353)

[9] "Daß port ist von menschen händen bey mans denken gemacht worden in daß diefe meer hinein, dergestalt: man hatt grosse stein von dem berg Monjuy in daß meer hinein gesencket, so haufecht auf einanderen, biß sie über daß meer hinauß gesehen haben, ettwan 20 schu hoch; demnach hatt man es mitt kalch unndt sandt wie ein kitt überschittet, ettwan 12 schritt breit unndt 500 schritt lang unndt gradt in das diefe meer hinaus, unndt ist sich zeverwunderen, daß daß meer gleich von dem gestaad auß so dieff ist, dahär daß port desto sicherer unndt besser." (Keiser 1968: 341)

4 Observations on Space Before the Great Extension in the 19th Century

When the travelers describe the structure of the city as a whole, they continue the tradition of the formal basic pattern by dividing the city into two parts: into old and new or, in other words, into a "ville vieille" and a "ville neuve". With this division, it has always been about the visible architectural structure, which refers to the two parts as being divided by the inner city wall—which in 1775 was transformed into a wide promenade (*las Ramblas*). Only rarely are other possible divisions mentioned by the travelers, but the one exception was made by Joseph Townsend (1739–1816), a British physician and geologist, when he showed more interest in the administrative districts and parishes than in the basic architectural structure.[10]

Within the field of urban forms, there are further variations. A French traveler at the beginning of the 18th century maintained that the capital city of Laletans (*Lalétains*) had a square shape. For his contemporaries he described the shape as being halfway between a square and a rectangle: "d'une forme entre la quarrée & l'ovale, & grande à-peu-près comme Naples" (Alvarez de Colmenar 1715: 600–601). Henry Swinburne, who visited the city in November 1775, described its shape as "almost circular, the Roman town being on the highest ground in the center of the new one" (Swinburne 1787: 24). Similarly, for a German traveler, it is once again round (Volkmann 1789: 352); another German describes it as a "Halbzirkel" (semi-circle), reminiscent of Platter's half-moon shape (Delius 1834: 28).

The attribution "ville neuve" or "Neustadt" (new town), however, no longer refers solely to the Raval in the second half of the 18th century, but also to the newly established settlement around the harbor. According to Giuseppe Baretti (1719–1789), a writer and the Secretary of the Royal Academy of Arts, who visited Barcelona in 1760, the city can be categorized in this way: "Les uns l'appellent *La ville neuve* d'autres *Barcelone la neuve*, & même *Barceloneta*." (Baretti 1777: 85). Barceloneta was presented as the work of the Marquis de la Mina,[11] the Governor of Catalonia, who had spared himself no trouble in order to support the enterprise. Up to that time, there were already 3,000 inhabitants living there, including some merchants (Baretti 1777: 87). Johann Jacob Volkmann (1732–1803) still upheld the bi-partite division of the city into old and new, but from 1752, he added that there was still "another little town called Barceloneta",

[10] "Barcelona may be considered as divided either into districts or into parishes; the former being five, the latter eight, including the cathedral." (Townsend 1792: 134).

[11] Jaime de Guzmán-Dávalos y Spínola, marqués de la Mina (1690–1767) (de Vicente Algueró 1984).

which in 1775 already comprised 2,000 houses and 10,000 inhabitants.[12] He claimed that it was "an elongated rectangle" and had about "20 fine, straight, wide streets" (ibid.). For Townsend Barceloneta is a suburb,[13] and the same applies to Delius, a man from northern Germany: a suburb on a peninsula (Delius 1834: 31–32); for Charles Didier, a Frenchman, it is more than a suburb, but rather "a town within a city", inhabited mainly by seamen and merchants.[14] Pierre Chantreau (1741–1808), a French linguist and publicist, suggested for the first time that there were three towns instead of the frequently quoted "2 + 1 structure"—a new town, an old town (referring to the former Roman town), and Barceloneta, situated by the harbor.[15] Loning (1844), a German army officer, also adopts the three-town model. However, the most extraordinary description was made by a French traveler called Cornille. For him Barcelona was a "ville bâtarde", a city of opposites or a town that is neither one thing nor another: here—the merchants quarter, there—the financiers and speculators; here—the Spanish part, there—the French quarter, a meeting point between North and South, where one can at the same time feel the wind of the Pyrenees and the wind of Andalusia (Cornille 1836: 227–228).

Most travelers take note of the harbor. Unlike the 16th century, the harbor is seen from then on to be wide, safe, deep, well laid out, and sometimes even as beautiful or convenient. Many travelers took an interest in the infrastructure of the harbor: the mole, the storehouses, the machines used for loading and unloading, the quay and the lighthouse. It seemed to be generally agreed that the harbor made a contribution to the commercial development of the city. Many travelers go so far as taking a register of the number of exports and imports, thus revealing the wide-ranging economic networks connected to the city.

However, the Citadel (1714–1718) was less a centre of interest; nonetheless, when it is referred to by the foreign travelers, it usually becomes a subject of

---

[12] "Er legte, weil die Stadt wegen der Festungswerke nicht vergrößert werden konnte, auf der in die See gehenden Strecke Landes, welche den Hafen formirt, 1,752 eine neue kleine Stadt Barceloneta an, welche 1775 schon aus mehr, als 2,000 Häusern und gegen 10,000 Einwohnern bestand. Sie macht ein längliches Viereck aus, und hat ohngefähr 20 schöne gerade breite Gassen." (Volkmann 1789: 354, translation by John Gledhill).

[13] This is taken from a supplement to the French translation (Townsend 1809): "*Barcelonetta*, ou *Petit Barcelone*, est une espèce de faubourg de Barcelone, situé en dehors de la porte de la Mer en formant un des côtés du port. [...] On estime que ce faubourg contient 13,000 âmes."

[14] "Barcelonette est plus qu'un faubourg, c'est une ville dans la ville; [...]. Barcelonette est une construction entièrement moderne; elle ne remonte pas au-delà du dernier siècle [...]. Il y règne une grande activité; c'est là qu'on fabrique les ancres, les voiles, les cordages, tout ce qui concerne la navigation, et l'on y construit même des bâtiments marchands de toutes grandeurs." (Didier 1841: 33–34)

[15] "Barcelonne [...] est sur un amphithéâtre de forme circulaire [...]. Elle renferme trois villes en une; la moderne, l'ancienne et *Barcelonette*, où est le port: la moderne en forme presque la totalité et enclave l'ancienne, dont les murs subsistent encore très-intacts dans beaucoup d'endroits; on les attribue aux Romains, parce que les Espagnols, très-ignorans en architecture et en antiques, leur attribuent tout." (Chantreau 1792: 203–204)

admiration.[16] For a traveler—who was probably a Catalonian—and who wrote under the pseudonym Poco Mas, the Citadel was "a small fortified town within the city" (Mas 1845: 372). He reports that one can read in the old chronicles that 600 houses, a church and three convents had to be pulled down in order to build this fortress. The first attempt at demolition—in the context of the 1841 People's Uprising—is recorded in the same year by Alfred von Bergh: at that moment, 1,100 people were working on the demolition of the Citadel (von Bergh 1841: 109). Towards the middle of the 19th century, the Citadel is no longer just politically and militarily in the wrong place at the wrong time, it is also taking up room, which could be put to good use by the city's population. Mas summed up the situation as follows: "I believe to be, that the population, the working population especially, of Barcelona has increased to such an extent that, being a fortified city, the people are now cramped for room." (Mas 1845: 373) Other travelers during 1840s also were aware of the necessity of expanding the city. August Ludwig von Rochau, a German publicist, goes so far as to attribute the cause of the political unrest to the inadequate spatial provision within the city. According to him, the houses are full right up to the rooftops, the traffic is threatening to come to a standstill in the narrow streets, rents are outrageously high, and there is no room for any commercial growth.[17]

[16] Chantreau mentions that it was built as a defence against the Cataloneans without, however, referring to its position: "La citadelle a six bastions qu'on eût pu réduire à deux, car ils en imposent plutôt aux Catalans mêmes qu'à l'ennemi, qui ne perdroit pas son temps à attaquer cette citadelle pour se rendre maître de la ville, parce qu'elle est dans un fond que la ville la domine, et qu'une esplanade immense l'en sépare." (Chantreau 1792: 70)

[17] "Auf der entgegengesetzten Seite der Stadt, in der Nähe des Thores, welches nach Gracia führt, ist eine ähnliche Bresche vorhanden, an die man bis jetzt noch gar nicht Hand gelegt hat, da es im Plane ist, die Stadt nach dieser Seite hin zu erweitern. Die Mauern von Barcelona sind jedenfalls zu eng geworden für die mit jedem Jahr wachsende Bevölkerung, und diese würde sich wahrscheinlich weniger ungeberdig zeigen als in den letzten Jahren, wenn man ihr mehr Raum zum Athmen und mehr Raum und Freiheit zur Bewegung gäbe. Das allgemeine Verlangen, das gebieterische Bedürfniß der materiellen Raumgewinnung ist ohne Zweifel eine sehr wirksame Mitursache der heftigen Volksbewegungen gewesen, durch welche Barcelona in dieser Zeit erschüttert ward. Die hohen Häuser sind bis an die Dächer vollgepfropft mit Menschen, der lebenskräftige Verkehr droht in den schmalen Gassen der Stadt zu ersticken, die Miethpreise stehen übertrieben hoch, und für neue gewerbliche Anlagen ist es unmöglich, Platz zu gewinnen, so daß z.B. die Zahl der jetzt in der Stadt arbeitenden Dampfmaschinen (etwa vierzig) dem Beschluß der städtischen Verwaltung zufolge nicht überschritten werden soll. Ein beträchtlicher Theil der Barceloneser Industrie hat sich nach Gracia hinausgeflüchtet, aber die Stadt sieht in solchen Auswanderungen natürlicherweise nur Verluste, die sie mit Unmut duldet." (von Rochau 1847: 86–87)

5 Projects Planned, But Not Realized

Although the projects which had not been carried out obviously cannot be viewed, planned projects which had been executed halfway or even not at all are still part and parcel of the city's planning and transformations. A brief glance in this direction can, on the one hand, highlight the difficulties—whether political, financial or technical—the actors sometimes had to contend with; on the other hand, this aspect shows that even imagined urban spaces need to be incorporated into a spatially oriented history of the city. This point is supported by the fact that some contemporary observers were also fascinated by this aspect.

In connection with Barceloneta, Baretti argued that, in the foreground of this site, an alternative was possible. Even before this time, there were plans to demolish the fortifications in order to extend the city. According to Baretti in his 76th letter, the governor had rejected this idea and had decided in favor of this site for the "new town" (Baretti 1777: 85). In November 1775, when Swinburne was in Barcelona, he had come across a project to build a canal, which would be controlled by a group of foreign investors: "Some years ago a company of Dutch and English adventurers offered to bring the river into the port by means of a canal, if government would allow them a free importation for ten years. This project might have cleared away the sand-banks, but might also have given a fatal check to the infant manufactures of the country, for which reason the proposal was rejected." (Swinburne 1787: 25) Volkmann, who had received Swinburne's report, took an active interest in this project, but invoked the same reasons for its rejection (Volkmann 1789: 353). Johann Gottlob von Quandt (1787–1859), an art historian, writes in October 1846, that people are at the moment busy with plans to move the harbor to another location; however, obstacles from all sides arose preventing the execution of this plan (von Quandt 1850: 22).

6 Observations on Space in the Context of the Great Extension Plan (ca. 1850–1870)

Around the middle of this century, travelers observing the city were often fascinated by the topic of the city's extension. Urban spatial movements and alterations now became the central focus of their interest. Less is written about the shape of the city: Alexander Ziegler suggests that the shape of Barceloneta is that of a right-angled triangle (Ziegler 1852: 95). The bi-partite division into old and new still appears now and then, but the city's earlier shape gradually dissipated so as to become hardly recognizable. According to Friedrich Wilhelm Hackländer (1816–1877), a writer, old Barcelona is on the Montjuich side whereas "new Barcelona" is situated in the direction of the Citadel because here it is more splendid and livelier and because the shops, stores and vaults are all crammed next to one another, and this is the mark of an up and coming city (Hackländer 2006:

107–108). Baumstark, who had read many travelogues, questions the assumed bipartite division of the city into old and new. To him the division was "keineswegs besonders schroff" (in no way clear cut) (Baumstark 1869: 20). Obviously, the Rambla emphasizes the division of the city into two unequal halves. The waning interest in the shapes (circle, rectangle or half-moon) is evidenced by the fact that the city wall at this time was losing its defining inner/outer demarcation. Its demolition began to take place in 1854 (Busquets 2004: 122–124), and thus the former separated areas were able to merge more and more into each another. Gracia, a former suburb situated to the north of the city, had become a part of the main city according to an English traveler writing in 1868 (Pemberton 1868: 326).

The travelogues written in the middle of the 19th century show less interest in the geometric shapes than in the movements within the city, and indeed not only in popular movements (von Rochau 1847: 87; Heinzelmann 1851: 35), in which the people expressed their dissatisfaction but also in urban spatial changes (Hackländer 2006: 100–140). Von Quandt maintains: "Barcellona ward mehrmals vergrössert und verschönert" (Barcelona was extended and improved many times) (von Quandt 1850: 13), which, in the following passage, no longer refers to the changes since antiquity, but rather to those of the previous century: the construction of Barceloneta, the renovations connected with the destruction from the revolutionary period and the widening of calle (street) Fernando VII. Valeska Voigtel-Bolgiani (1830–1876), publishing under his pseudonym, Arthur Stahl, offered the following glance into the future: "Only after some years will Barcelona attain its full splendor. The fortified walls which had fallen down in 1843 and then leveled down just like those in Vienna, have given the cramped city unlimited space for expansion. The new buildings which have been designed according to a set plan will become a most impressive sight."[18]

Most of the travelers at this period of time moved on foot through the city, or, in other words, they walked from one place to another or one sight to another, and either presented their own impressions or reproduced what they had read. The only travelers whose routes are difficult to follow are those who enjoyed only a brief stay in the city or who failed stylistically to write a travelogue based on experience. These writers tend merely to offer their general impressions and a few facts. A few city walkers deliberately climbed a hill (usually the Montjuich) in order to gain a panoramic view of the city, which in the middle of the 19th century had between 150,000 and 200,000 inhabitants and, as a result, it was difficult for travelers on flat ground to find their way about without a map. Friedrich Heinzelmann walked up onto the Montjuich and noted five points between which Barcelona was "eingekeilt" (wedged in): Fort Montjuich, Fort Atarazanes at the foot of a mountain, the (renovated) Citadel together with Fort Pio and Fort

[18] In the original German version: "Zu ihrer vollen Schönheit wird sich Barcellona erst nach einigen Jahren erheben. Die Festungsmauern, seit 1843 gefallen und gleich denen von Wien abgetragen, haben der eingeengten Stadt unbeschränkten Raum zur Ausdehnung gegeben. Die Neubauten, nach einem gewissen Plan ausgeführt, werden einen höchst imposanten Anblick gewähren." (Stahl 1866: 215, translation by John Gledhill).

Canaletas (Heinzelmann 1851: 34). He linked these external fixed points to the city's narrow streets, in which any army trying to take over any part of the city would be defeated. Franz Lorinser (1821–1893), a clergyman, also begins his description with a panoramic view, in order to be better able to compare Barcelona with other beautiful Spanish cities such as Murcia, Granada and Toledo. He describes Barcelona from the bird's eye point of view for the reader and, in conclusion, he regrets the fact on account of such a wonderful view that the omnibus due to take him back to the Fonda (his lodging) is not able to fly him back like a bird. This means that the raised position was, in fact, pure fiction. However, according to so Lorinser, it is possible to have an idea of the view depicted by the clergyman if one climbs up to Fort Montjuich.[19]

7 Summary and Conclusions

At the beginning of the 19th century, an anonymous traveler, in Barcelona writes that one can identify the various "Zeitpuncte" (key points in time) of its history by examining the remains of the city wall.[20] An analytical exposure of the spatial layers together with their tendency to merge into one another by no means conveys a picture of history progressing chronologically in a straight line, but creates more an image of the interwoven complexity of various periods of time and epochs, of changing sites, of expansions, which incorporate diverse parts of the old within the new. A traveler can be compared to an archaeologist trying on the one hand to

[19] "Man denke sich einen Halbkreis mäßig hoher, zierlich geformter blauer, felsiger Berge, die in fruchtbaren Hügeln allmählich gegen das Meer hin sich verflachen und nur im Süden einen mächtigen, steil abfallenden Felscoloß in's Meer vorgeschoben,—und im Schooße dieses Halbgürtels eine bis unmittelbar an das Gestade sich ausdehnende Ebene, die von einer gewaltigen Häusermasse bedeckt ist, welche durch eine breite, mit Bäumen bepflanzte Straße, die am Hafen beginnt und am westlichen Thore endigt, in zwei ungleiche Hälften getheilt wird, und aus der eine Menge sich sehr ähnelnder Thürme, die alle oben platt abgeschnitten sind, emportauchen, und daneben die unermeßliche, stahlblaue Fläche des Meeres, in welche eine mit weißen Häusern bedeckte, sich krümmende Landzunge (Barceloneta) hinausläuft,—und gegen Norden eine langgestreckte Küste, an der wie blendende Edelsteine die beiden Städtchen Badalona und Matarò liegen,—und über Alles eine prächtige Beleuchtung ausgegossen,—und, um das Bild zu vollenden, im Vordergrunde wilde Felsparthieen und colossale Agaven, und man wird ungefähr eine Vorstellung des Anblicks erhalten, der sich vom Fort Montjuich, wenn man das Gesicht nach Norden wendet, darbietet. Das wäre Barcelona aus der Vogelperspektive. Zu solchem Fluge vermochte freilich der Omnibus, der uns vom Bahnhofe in die Fonda führte, sich nicht zu erheben." (Lorinser 1855: 58–59)

[20] "Noch erkennt man an Barcelona's verschiedenen Einlassungen die unterschiedenen Zeitpuncte seiner Geschichte. Anfänglich lag sie auf einer Höhe unfern dem Meere, fast wo jetzt der Mittelpunct des Ortes ist. Noch sind viele Thore dieser alten Ringmauer vorhanden. So wie unter Arragoniens Königen der Wohlstand Barcelona's zunahm, wurden auch dessen Ringmauern verlängert und vergrößert. Der gegenwärtige Umfang ward der Stadt nach den Kriegen in der Mitte des vorigen Jahrhunderts." (T. 1817: 114)

uncover the different layers of history and on the other hand having to decide what to make of remains he has uncovered. However, whatever the case, his excavations bring materializations of time in space to the light of day, which can be described as a chronotopos as defined by Michail Bachtin, the Russian cultural scientist (Bachtin 2008). Only in this way will the changing space–time configurations of a city become recognizable. A differentiated consideration of perceived times in space can be transferred analogically to the perceived spatialities. Just as linear chronologies are inadequate for the analysis of past societies, three-dimensional spatialities fall short here because the spatial praxis of the actors would be inadequately described.

In this context, if the question is once again posed as to how one can map historical urban spaces, it first hast to be decided as to exactly what kind of "space" do we want to map. It is absolutely obvious that the space of geographers, cartographers, city planners, city fathers or social-geographers is not the same space as that recorded, analyzed and covered by everyday users of space such as, in our case, the travelers themselves.

Before the actual mapping of historical urban spaces can take place, methodology demands that the following steps need to be observed and the following questions to be cleared up:

- Which spaces (urban space types—such as streets, town squares, buildings—shapes, hierarchies, relations, distances etc.) are to be taken into consideration and mapped?
- What are the perspectives (including fictive ones) taken by the observers? What kinds of instruments or media do they use to enhance their perception of space? In what ways do they find their orientation in urban space?
- Which spatial modalities are the observers aware of or, alternatively, which do they deem to be worth recording? Obviously this does not just concern the static space of the city, but also both changes of space and bodily movements. In this way, spaces are often created, which are then translated by the writer into a narrative.

From this, the general direction for social geography with regard to historical urban research will acquire a critical and necessary opening up of perspective which involves the inclusion of perceptions, movements and transformations.

The various modalities need now at the very latest to be taken into account, in which temporality in urban space appears or is created. Alterations in urban space do not progress in the mode of time moving forward in a linear direction. The descriptions of the transformations show how far they are subject to narrative constructions or intertextual dependencies. Finally, the observers need to be aware of the multi-layering within space and the co-presence of several time layers in space.

With all this in mind, now would be the time to map movements in urban space, perceptions of urban spatial transformations or urban spatial hierarchies together with their changes.

References

Alvarez de Colmenar DJ (1715/1707) Les délices de l'Espagne et du Portugal, vol 4. Pierre vander Aa, Leiden
Bachtin MM (2008/1974) Chronotopos. Suhrkamp, Frankfurt
Baretti GM (1777) Voyage de Londres à Gênes: passant par l'Angleterre, le Portugal, l'Espagne, et la France, vol 4. Marc-Michel Rey, Amsterdam
Baumstark R (1869) Mein Ausflug nach Spanien im Frühling 1867. Georg Joseph Manz, Regensburg
Busquets J (2004) Barcelona: La construcción urbanística de una cuidad compacta. Ediciones del Serbal, Barcelona
Casellas A (2009) Barcelona's urban landscape: the historical making of a tourist product. J Urban Hist 35:815–832
Chantreau PN (1792) Lettres écrites de Barcelonne à un zélateur de la liberté, qui voyage en Allemagne. Buisson, Paris
Chmel J (ed) (1846) Herberstein's Gesandtschafts-Reise nach Spanien 1519. Peter Rohrmann, Wien
Conley T (1997) The self-made map: cartographic writing in early modern France. University of Minnesota Press, Minneapolis
Cornille H (1836) Souvenirs d'Espagne, vol 2. Arthus Bertrand, Paris
de Certeau M (1990) L'invention du quotidien, vol 1, Arts de faire. Seuil, Paris
de Vicente Algueró FJ (1984) El Marqués de la Mina, de militar profesional a ilustrado peroférico. In: Primer Congrés d'Història Moderna de Catalunya, vol 2. Facultat de Geografia i Història, Barcelona, pp 89–100
Delius E (1834) Wanderungen eines jungen Norddeutschen durch Portugal, Spanien und Nord-Amerika: in den Jahren 1827–1831, vol 2. Herold, Hamburg
Didier C (1841) Une année en Espagne, vol 1. de Dumont, Paris
Füssel M (2013) Tote Orte und gelebte Räume: Zur Raumtheorie von Michel de Certeau S.J. Historical social research 38.3, pp 22–39
Hackländer FW (2006/1855) Ein Winter in Spanien, vol 1, Breuer T (ed) http://www.fw-hacklaender.de. Accessed 14 Feb 2013
Heinzelmann F (1851) Reisebilder und Skizzen aus der pyrenäischen Halbinsel nebst Blicken auf die Länder des mejicanischen Golfes und Californien. Friedrich Fleischer, Leipzig
Keiser R (ed) (1968) Thomas Platter d.J.: Beschreibung der Reisen durch Frankreich, Spanien, England und die Niederlande 1595–1600, vol 1. Schwabe, Stuttgart et al
Knowles AK, Hillier A (eds) (2008) Placing history: how maps, spatial data, and GIS are Changing Historical Scholarship. Digital supplement. ESRI Press, Redlands
Lefebvre H (ed) (2000/1974) La production de l'espace, 4th edn. Anthropos, Paris
Lestringant F (2012) Die Erfindung des Raums. Kartographie, Fiktion und Alterität in der Literatur der Renaissance. Erfurter Mercator-Vorlesungen. Transcript, Bielefeld
Loning A (1844) Das Spanische Volk in seinen Ständen, Sitten und Gebräuchen, mit Episoden aus dem carlistischen Erbfolgekriege nach eigner Anschauung und Quellen. Hahn'sche Hofbuchhandlung, Hannover
Lorinser F (1855) Reiseskizzen aus Spanien. Schilderungen und Eindrücke von Land und Leuten, vol 2. Joseph Manz, Regensburg
Mas P (1845) Scenes and adventures in Spain from 1835 to 1840, vol 2. Richard Bentley, London
Münzer J (1991) Viaje por España y Portugal (1494–1495). Edición Polifemo, Madrid
Pemberton H (1868) A winter tour in Spain. Tinsley Brothers, London
Rau S (2011) Zeit-Räume, Parcours und Karte: Die Raum-Erkundungen der Reisenden in frühneuzeitlichen Großstädten. In: Tiller E, Mayer CO (eds) RaumErkundungen: Einblicke und Ausblicke. Winter, Heidelberg, pp 155–180
Rau S (2013) Räume: Konzepte—Wahrnehmungen—Nutzungen. Campus, Frankfurt

Reimann V (1996) Neue Zentralitäten: Die europäische Metropole im Strukturwandel; das Beispiel Barcelona. Cuvillier, Göttingen
Roche D (2006) Les livres de voyage à l'époque moderne XVIe–XVIIIe siècle. Revue de la Bibliothèque nationale de France 22:5–13
Rodríguez-Lores J (1980) Ildefonso Cerdá. 1980. Die Wissenschaft des Städtebaues und der Bebauungsplan von Barcelona (1859). In: Fehl G, Rodríguez-Lores J (eds) Städtebau um die Jahrhundertwende. Materialien zur Entstehung der Disziplin Städtebau. Deutscher Gemeindeverlag and Kohlhammer, Cologne, pp 36–83
Sabaté Bel J (1985) Die Stadterweiterung von Barcelona zwischen 1859 und 1891. Zur Kontrollfunktion der Bauordnungen bei der Verwirklichung des Stadterweiterungsplanes von Ildefonso Cerdá. In: Rodríguez-Lores J, Fehl G (eds) Städtebaureform 1865–1900: von Licht, Luft und Ordnung in der Stadt der Gründerzeit. Christians, Hamburg, pp 423–446
Santa-Maria Batlló G (2009) Decidir la ciutat futura. Barcelona 1859. Museu d'Història de Barcelona, Barcelona
Stahl A (1866) Spanien: Reiseblätter, vol 2. Otto Wigand, Leipzig
Swinburne H (1787) Travels through Spain, in the years 1775 and 1776, vol 1. J. Davis, London
T DGW (1817) Spanien, oder Sitten, Gebräuche, Trachten und Denkmäler der Spanier, vol. 1. Hartleben, Pesth
Torres i Capell M (1985) Inicis de la urbanística municipal de Barcelona. Ajuntament de Barcelona, Barcelona
Townsend J (1792) A journey through Spain in the years 1786 and 1787, vol 1. C. Dilly, London
Townsend J (1809) Voyage en Espagne dans les années 1786 and 1787, vol 1. Dentu, Paris
Volkmann JJ (1789) Neueste Reisen durch Spanien vorzüglich in Ansehung der Künste, Handlung, Oekonomie, und Manufakturen auß den besten Nachrichten und neueren Schriften zusammengetragen. Caspar Fritsch, Leipzig
von Bergh A (1841) Letzte Reisebriefe über Portugal und Spanien. Decker, Berlin
von Quandt JG (1850) Beobachtungen und Phantasien über Menschen, Natur und Kunst auf einer Reise durch Spanien. C. L. Hirschfeld, Leipzig
von Rochau AL (1847) Reiseleben in Südfrankreich und Spanien, vol 1. Cotta, Stuttgart
Welsch H (1658) Wahrhafftige Reiß-Beschreibung aus eigener Erfahrung….Johann Weyrich Rößlin, Stuttgart
Ziegler A (1852) Reise in Spanien: mit Berücksichtigung der national-ökonomischen Interessen, vol 1. Friedrich Fleischer, Leipzig
Zimmermann C (2000) Die Zeit der Metropolen: Urbanisierung und Großstadtentwicklung. Fischer, Frankfurt

Rethinking Social Relations: Towards a Different Phenomenology of Places

Stephen Read

Abstract So-called phenomenological approaches to the understanding of social and spatial relations usually deal with these in terms of 'mental space', 'existential space', 'social space' and so on. These modes of space are regarded as 'subjective', 'soft' and short on the 'hard' mathematical, geometric or objective properties that give spatial analysis a rigorous analytical capability. I argue here that this misrepresents and misunderstands a central principle of phenomenology and overlooks phenomenology's potential to *objectively* map us in our world. In its essence phenomenology is founded on the relation of *intentionality*. It is not necessarily about an interior mentality at all but about a subject-object relation in the world. The model that says there is an interior subjective or imaginative realm on the one hand and an exterior objective, physical or real one on the other, between which relations must be established for human knowledge or action to be produced, is replaced by one in which a perfectly real subject at one end of an intentional relation is connected to a perfectly real object at the other. A different phenomenology of places would be about how these relations between subjects and objects are structured and intentional knowledge and action mediated in the world. It would be about the *environmental* relation where the notion of 'environment' is captured in the relations between intentional subjects and the objects of those subjects' attention and intention. I argue that this is eventually about how we order and construct human 'worlds' technologically and spatially so that we may effectively inhabit and use them. These 'worlds' exist as whole networks of subjects and objects, in part-whole, mutually constitutive, relations with 'worlds'. The translation of the intentional relation into geography and urbanism involves us in an historical process of the construction of metageographical structures through which subjects establish and order their knowledge of and practices in the world. Enclosures, divisions and connections made by us in the world have shaped these structures and established the geographical and urban frames of our lives. This requires us to understand the human

S. Read (✉)
Department of Urbanism, TU Delft, Delft, The Netherlands
e-mail: s.a.read@tudelft.nl

S. Rau and E. Schönherr (eds.), *Mapping Spatial Relations, Their Perceptions and Dynamics*, Lecture Notes in Geoinformation and Cartography,
DOI: 10.1007/978-3-319-00993-3_8,

world as an historical construction, an *anthroposphere*, of regions and places, as *equipment* for framing our knowledge of the world and our local and translocal actions in it. I start by looking critically at social relations as these are imagined today, finding their origins in an Enlightenment metaphysics which bifurcates nature into mental and corporeal realms, and suggest an alternative founded in this reassessment of phenomenology. This alternative centres our attention on the anthroposphere as a construction, and a topological 'structure of places', organised as a layering of places and infrastructural 'grids' into a set of normative 'levels' which have a metageographical, intelligibility-giving and practice-defining character. 'Structures of places', 'grids' and 'levels' are perfectly objective and mappable and are proposed as the foundation of a new phenomenological urban and geographical model.

1 Approaching Relations Anew

The technical disciplines which map human or social relations generally formalise these as models and in terms of the spaces these models incorporate. It is this kind of mapping and these kinds of mappers I am addressing. Phenomenology has most often been associated here with a 'soft' subjectivity and with the perspective of individual experience. Phenomenological mappings are therefore often of individual and subjective 'perceptions' of an objective world. With this emphasis on the individual subject, the perspective extends also into 'individual choice' approaches where, in 'time geography' (Hägerstrand 1978), for example, structure is produced by individuals with personal 'time budgets' making individual choices about movement and other behaviours. At the same time, cognitive approaches, like that of 'mental mapping' (Lynch 1960), see the individual perceiving subject producing mental maps represented in the individual interior mind. Phenomenology has indeed shared the belief that humans are 'cognitive' subjects (Varela 1996), internally processing representations of the world, rather than being simply situated in-the-world, and although most phenomenologists after Husserl have understood the subject as being in a direct relation with a material environment, the multiplicity and variability of individual and subjective spaces has still for the most part been assumed to be an effect of the 'error-prone' nature of subjective perception and representation when faced with a definite singular objectivity. This 'Cartesian doubt' has also been behind questions about the reliability of phenomenological descriptions and interpretations (Seamon 2000). The assumption has been that the mind is a rather inexact, even faulty, computer, and the basis of this assumption is an Enlightenment model of 'mind' incorporating a division between a personal and error-prone subjectivity and perception and an objective reality and certainty.

Both the association of subjectivity with the individual and the interiority of subjective processes will be treated here as problematic. The question of the mediation of perception and intention should be treated in reality and in the world, and not collapsed, as I will outline, to one of crossing a 'gap' between subjectivity and objectivity or mind and world. The map itself also needs to be reconsidered. Normally considered in orthographic projection (in time geography with a time axis added), the map has been associated with the objective side of the subjectivity-objectivity division and has embedded problematic assumptions about the nature of our perspective on and our knowledge of the world. It has also been associated with an absolute space of cartographic or geodesic coordinates and a Euclidian geometry and distance that apply universally. The map I will propose can be represented orthographically for visualisation and analytical purposes, but will contain objects, subjects and places organised in their own humanly-constructed spaces that embed normative geographical characters and scales. The coordinate geometry and distances of the map itself are not assumed to be 'objective' in relation to these spaces and are not determining of these spaces.

I will begin by reviewing and critiquing social relations as we most often think of them today. This will necessitate following a discussion and critique of Enlightenment metaphysics, in terms of the relations between subject and object, and mind and world, before building a proposal that our human-world relations are an objective factor of the world itself understood in terms of these relations. I will be addressing the question of an alternative to the division of existence into mind and world, firstly by reviewing phenomenology's well-known role in a critique of artificial intelligence and then by way of a reinterpretation of phenomenology in 'non-subjective' terms, and in relation to technology and equipment, that was highlighted in an hermeneutical philosophy of science. The alternative phenomenological view I will present is one that emerges first by restating the foundation of phenomenology in the intentional relation and by posing subject and object as the two objective poles of this relation, and then by situating this relation in the world and finding mediation in the way this relation is reliably enacted in the world. This reliable enaction takes place through the equipment and artefacts we surround ourselves with. I will restate this view also in terms of an urban model consisting of patterns of objects, places and network infrastructures constructed in the world. These patterns will incorporate a 'structure of places' together with the infrastructure technologies that support these.

This alternative non-mentalist, non-subjectivist[1] 'phenomenology of places' will problematise our conventional view of subjectivity and mind, and reveal the roles of objects, technologies and construction, and our own nature as emplaced creatures, enabled and constrained by this emplacement. In short then, I aim to rematerialise and 'complexify' social relations, bringing them back into the world. In doing this I hope at the same time to say something about technique and construction as qualities and characters of the human and the social. We should

[1] Patrick Heelan, to whom I will refer later, used the word 'non-objective'. The point is that the subjective–objective dichotomy is rendered superfluous in a position which understands us as actors and constructors of a sociotechnical world rather than subjects reflecting from an interior subjectivity onto an exterior objective world.

again be able to map human and social things in an objective way, but in a way which also finds subject positions and social-technical conditions and constructions back in our objective mappings.

2 Taking the World Out of Social Relations

As researchers and builders of models and theories, our intuitions are powerfully formed by the unstated philosophies and metaphysics we work with. This is not something we can circumvent; such philosophical or metaphysical 'backgrounds' are a condition of our working effectively in 'communities' of 'like-thinkers', but they are also a constraint on what may be possible or allowable to think in these 'communities'. Thomas Kuhn was one of the first to highlight this social and 'practice' dimension to knowledge (Kuhn 1962). This means our intuitions are not always the best guide as we go forward and insights from outside the 'community' can reveal what is unquestioned in the 'community' and open new ways to progress. In what is for many the most intuitive way of modelling social relations, ties of interaction between humans as 'social agents', and the structural effects of these are analysed. Social network theory (Wasserman and Faust 1997; Scott 2000) takes social networks to consist of social agents and the ties between them and not much more than this. Starting from this parsimonious set up social network theorists can begin to think how individuals can come together to create enduring, functioning societies. Georg Simmel had suggested that social reality consists of interrelated clusters in which smaller groups or units are built up into larger ones which then overlap with one another (Simmel 1955; 130–135). Emile Durkheim had argued that human societies are made up of interrelated individuals and that the reasons for social regularities are to be found in the structure of 'social environments' rather than in the individuals (Durkheim 1951). One of social network theory's founding fathers envisioned it as a kind of physics, complete with its own social 'atoms' and 'laws of social gravitation' (Moreno 1934) and social networks provided a way of making tangible the sorts of structure Simmel and Durkheim conceived.

The basic structure created in social networks is the 'group' or 'community', so that a structural scaling is achieved and agency may transfer from the social agent to higher-order structures which may then 'know' and 'act' in their turn as sorts of super-individuals. However, the social network idea has failed to account for some of the things we might expect to learn from relations. In particular, while the theory accounts for some non-local structural effects of local ties, it has difficulties with understanding higher-order structure or *globality* as an active contextual dimension, or indeed any active relations beyond the direct relations of individuals or groups (Friedman 2007; Robertson 1992; Chapman 2004).[2] Social network

[2] The influence of globality, or the world, on our everyday sense of order or coherence is not captured in the aggregative social networks model—which needs the larger material frameworks I will be suggesting here to convincingly stabilise social relations.

theory has been criticised for its abstraction, its 'reification' of 'strong ties', and its mechanical understanding of the individual and of social accretion. How higher-order 'communities' are durably maintained along with their 'moral orders' (Latour 1992) remains an open question and I will return to this later.

The extent of the abstraction is evident from the fact that no material world appears as a relevant constituent or even background to these relations. Society is a 'relational abstraction' consisting of nothing more than the matrix of social relations. The 'social world' or 'social environment' is simply this abstraction, and we are not asked to consider how it may be possible to imagine relations at all without the mediation of a material world. A tacit assumption concerning the autonomous active nature of the subject is supported and maintained, while what is object is assumed to be either irrelevant or subject to the transparent influence or control of its active opposite. In fact the assumption is that we may abstract the social world away from its material supports as easily as we separate mental and corporeal realms. We imagine social relations inhabiting an autonomous 'immaterial' domain analogous to the mental, and social networks thoroughly internalise the assumption of the division of existence into active mental and passive corporeal realms and reproduce this metaphysics as a social model.

To see how this works we could revisit the foundations of our modern system of thought as this was formulated by Hume, Locke and Descartes. However, these metaphysics have already taken an interesting contemporary turn in artificial intelligence. Hume and Locke had conceived mind as a symbol-manipulating machine and AI researchers had simply applied this model as the digital computer became available to them, to build Enlightenment metaphysics into the computer, and embed, or so they thought, mind into computing machines. Underpinning AI therefore was the classic division of nature into *res extensa*—corporeal substance—and *res cogitans*—mental substance or consciousness. AI formalised relations between the poles of this division by way of a 'representational-computational' model. This model had 'minded-beings' 'interiorising' an immaterial representation of an 'exterior' material world which was then 'processed' in order to produce knowledge about and action in the world (Fodor 1981; Pylyshyn 1980). Perceptual 'input' was 'internalised' as manipulable symbols which could then be computed to produce 'output' in the form of abstract knowledge about or action in the world. In this scheme intelligent minds know about or act upon dumb matter, and mind is the active polarity while exterior reality remains acted upon. As the Heideggerian philosopher Hubert Dreyfus suggested, AI was "hard at work turning rationalist philosophy into a research programme" (Dreyfus 2007: 247).

3 Putting Mind in the Network

Questions about the abstraction of social networks have, however, also given rise to elaborations of this basic idea. In the 1960s Melvin Webber warned against "some deep-seated doctrine that seeks order in simple mappable patterns"

(Webber 1963: 54). He suggested our communities were becoming disentangled from the constraints of the models we used (including those of strong ties, place and distance) and highlighted other associational and institutional ties formed in modern social organisations and divisions of labour. His 'urban realms' were groups of people in 'community' with one another across geographical space and without being spatially delimited. He thought of these as cultural rather than territorial 'communities' sharing activities and knowledge 'without propinquity' (Webber 1963). Webber pointed at the same time to the ways new technologies were freeing the constraints of proximate face-to-face relations and creating "non-place urban realms" (Webber 1964).

Webber's social relations began to incorporate some of the new structures and constructions of modern society, but were no less 'immaterial' and abstract. Meanwhile 'space' was corporeal and becoming steadily less relevant and more threatened as embodied 'proximity' and 'face-to-face relations' (Boden and Molotch 1994). At the same time, a new factor, 'networks'—understood in terms of new communications technologies and high-speed mobility—was emerging. Cities were redrawn in these terms as dense clusters of no longer very relevant 'space', consisting of proximate social links, on the one hand, and 'networks', comprising 'abstract' and 'virtual' business, economic, cultural, associational and social links, on the other. The spatial disintegration or 'explosion' (Wright and Stewart 1972; Webber 1963) of the cities of the USA after the middle of the twentieth century, followed rapidly by those in Europe and other regions, was attributed to these new networks. The networks gave 'structure' to Webber's 'non-place urban realm' and 'community without propinquity' by mediating these through technological and especially 'microelectronic' networks. It was in these terms that technology entered 'social networks' as a medium through which human and social relations of 'networked individualism' (Wellman 2001) and a 'network society' (Castells 2000a) became possible.

This was hailed as no less than a revolution in social organisation and relations as, according to this view, technology established new structures of social relations and ushered in an 'Information Age' (Castells 2000a, b, 2004). The networks were seen as the armature of contemporary business and industrial production and of globalisation. They were also seen as the armature of a contemporary 'network society' "whose social structure is made of networks powered by microelectronics-based information and communication technologies" (Castells 1989). This is the foundation of the distinction made today between 'spaces of places' and 'spaces of flows' (Castells 1989) in which a dynamic contemporary business and society is mediated through an electronic 'space of flows', leaving an older 'space of places' of place-bound and embodied communities in a state of stasis and decline. A 'world' of sorts has entered the 'relational abstraction' with the network technologies, but the basic metaphysics remains intact, with machines and networks facilitating active but disembodied relations in a 'virtual' electronic realm. For Webber and many others, urban society has escaped the place-bound city and "the concept of 'urbanism' and the concept of 'city' are no longer coterminous" (Webber 1996). The pressing urban problem becomes the 'mediation' between a 'virtuality' of a 'space of flows' and a 'reality' of a 'space of places'.

4 Dismantling the Metaphysics of 'Virtuality'

The contemporary division of network 'virtuality' and place 'reality' is a corollary of a system of thought that has its origin in the seventeenth century. The continuing power and persistence of this system may owe as much though to its being embedded in influential research and development programmes as it does to the innate persuasiveness of the philosophies of Hume, Locke and Descartes. These research programmes have succeeded in building this system of thought not just into popular discourse and academic literatures and debates but also into the machines and networks through which we conduct a large part of our lives. A product of this is the 'naturalness' of the idea of a 'network society' as a 'virtual' 'relational abstraction' shaped in high-tech information networks today (Castells 2000a).

The products and contents of 'informational' networks are seen as a weightless, frictionless 'knowledge' or 'information' (see Read 2009). This maintains knowledge as immaterial and placelessly available, and seems to negate worldly mediation. In fact the mediation between autonomous mental and corporeal domains re-emerges in this model as a mediation between a non-place, dynamic and frictionless domain of a 'network society' (a 'space of flows') and a static, conservative and threatened domain of real places (a 'space of places') (Castells 1989). In the 'space of flows' and in our relation with machines, knowledge is not intrinsically or necessarily situated but exists in a placeless 'connected' network realm. When this knowledge is situated, it is usually presented to us in orthographic projection. The relation of knowledge to situation is made problematic again as the knowledge of the observer from orthographic perspective is conflated with knowledge on the ground. Philip Agre has noted how we see this clearly in computer games where the knowledge of the player is conflated with the knowledge of the 'agent' on the screen (Agre 1997).

However AI was not simply a channel through which Enlightenment metaphysics invaded our technics and artefacts; it was also the forum for an extended and continuing debate about the hidden metaphysics of this model and the search for an alternative. In this debate, Dreyfus started by suggesting that mind and world (or social actor and environment) are not separated in the first place and don't need to communicate across any mind-reality gap at all. For him knowledge and action should not be credited to the internally-knowing subject but rather to the situation of the subject in the world and the relation between subject and world. The key for Heidegger and for Dreyfus was that in bringing things to our knowledge or imagination, we incorporate them with our situated selves (*Dasein*) and begin relating to them directly. Things 'are' in a very important sense in the way we form a relation with them and take them into our lives and activities. Worldly things are disclosed to us in our relation to them and in the way they become something for us.

We don't relate, in other words, across a gap between us and a world, we are simply in it. Reality here is not accessed through interior representation but is practical, situated and *there* to us. Dreyfus' critique was to some extent provoked, and the argument fuelled and sustained, by a lack of progress in formalising intelligence within the prevailing paradigm. Dreyfus questioned the presumption that processes of mind involved the manipulation of abstract symbols according to sets of formal rules (Dreyfus 1965). He questioned also whether reality consists of atomic facts and things, and whether knowledge is even in principle formalisable in terms of universally applicable rules or laws. His alternative was a radical contextualism, shifting thinking away from an idealisation and internalisation of order and intelligence in mind or machine, and towards an in-the-world basis to order and reality. Cognition, understood as an individual and private, internal process, becomes redundant when the reasons for knowledge and action are situated externally and publicly in-the-world (Read 2008, 2013).

Dreyfus' intervention provoked some prominent AI researchers to participate in open-ended research in "psychology by reverse engineering" (Haugeland 1997). Rodney Brooks for example based his robotics on the idea that the world served as the sufficient 'representation' of itself in the domain of reality (Brooks 1997) and his machines were set off to simply learn on the go. Terry Winograd recognised that the sharing of coherent and meaningful knowledge or information depended on sharing a common context or frame of reference for that meaning. The representation-computation model had allocated meaning and significance to things through representation and language, but Winograd recognised we don't so much exchange information in communication as call to attention some aspect of the world the communicators already share (Winograd 1972). Part of that context of agreement is of course the world of everyday stuff we surround ourselves with.

Dreyfus emphasised the 'embodiment' of action and being in the world and the situatedness of knowledge. It's not us (or intelligent machines) who learn something about the world, but us and world who come together in significant conjunctures of knowledge or action and in a continuous process of relation and mutual adjustment. What Dreyfus' intervention in fact highlighted was our relations with respect to a world of things and both our and their integration with and dependence on this context or 'world' for practical 'being' or knowledge to emerge. As Don Ihde points out, this is less a matter of 'embodiment' and more a shift of the location of the 'subject' from a bodily interior to a directed relation with the world and with things (and people) in that world (Ihde 2010). From this point on the emphasis will be on the nature of that world, and in particular on its adjustment and preparedness for us and for our intentional knowledge and action.

5 Redefining Objectivity

This suggests that material things have meaningful existences in relation to us and our activities, and beyond our mental images and language, and that being out there in the world and amongst things is what is significant for subjectivity and social organisation. Others, including philosophers like Hannah Arendt, anthropologists like André Leroi-Gourhan, and sociologists like Bruno Latour, have developed this insight. Arendt sees us inhabiting a "world between things" (Arendt 1970), Leroi-Gourhan sees us making human worlds by 'exteriorising' a 'curtain of objects' (Leroi-Gourhan 1993), while Latour has argued that 'nonhumans' develop social relations just as much as humans do. He has suggested that humans alone do not and cannot hold our communities and societies together and that the 'missing mass' in our societies consists of nonhuman relations and the spatial organisation of objects and artefacts (Latour 1992). He emphasises an order 'networked' in arrangements of humans and nonhumans and the actions that join them up. Indeed, he argues that in these networks we find it difficult to separate the respective human, practice and technical parts and that these parts come together in 'negotiations' between people and things—particularly people and artefacts. With the reintegration of mind and world, we seem to arrive also at the reintegration of other polarities, like those of people and things and society and technology, and we could start thinking of these human-nonhuman arrangements as 'sociotechnical systems' (Hughes 1987).

These networks and systems are different to the ones we saw in 'network societies' in which advanced technology ends up appropriating the dynamic, creative role mind takes in the classic mind-world bifurcation. Latour's examples of objects and technologies—like that of the door-closer—testify to an active integration of humans and nonhumans that goes far beyond microelectronics. This suggests a very different metaphysics and we can begin to get a sense of what this metaphysics means from hermeneutical phenomenologists of science like Patrick Heelan, who looked at the hands-on practice of experimental science. Heelan pointed to the way the ostensibly subjective factor of the intentional human experimenter is 'hermeneutically shifted' in experimental practice into the ostensibly objective equipment of experiment. The observer is, according to Heelan, not outside the experiment, making notes from his or her subjective position about what is objectively going on. Equipment is not a passive 'container' of the experiment. Instead the equipment is designed and is being minutely manipulated, adjusted and maintained by the observer and other operatives, so that it becomes aligned with the intentionality of the observer and ends up producing the results intended (Heelan 1977). This is also the character of Latour's 'negotiations' between human and nonhuman factors: nonhuman material acts with the human actor in an 'actor-network' (Latour 2005). We could say that subjectivity, objectivity and practice are distributed in a closed network that has the character of the 'world' I will discuss later. Don Ihde calls this a 'material' or 'extended' hermeneutics—a directed 'negotiation' with things and processes until they end up doing what we require of them (Ihde 2009).

In this so-called 'technoscience' perspective acting and knowing escape the bounds of the subject as this is customarily defined, and creep out into the equipment, which is to some greater or lesser extent already prepared to do what is expected of it. Mediation here is by way of a 'subjectivity' escaped from the bounds of the subject and into worldly material, already prepared by engineers, builders and technicians. Mediation works through subject, object, practice in a networked and systematised way. Instead of imagining a society mediated in high-tech electronic spaces, and then imagining this high-tech as a corollary of an immaterial 'virtual' mind, we should start from the position that human relations with the world have long been mediated technologically, with technology understood here in a broad sense. Paul Edwards alerts us to the fact that the ubiquity and low-tech nature of most of the technologies that surround us hides from us their central roles in human knowledge and action (Edwards 2003). This suggests also that if networks and technologies have always defined what our subjectivities and societies are capable of, the high-tech networks and technologies of today may be just the latest telling of the same story. A more historical take on this may show also how the networks of today are by necessity tied to processes of technological innovation and substitution which sees mules and camel trains replaced by railways and trans-continental highways, horse-borne messengers replaced by postal systems and email and electronic messaging systems, and the carvel replaced by the container ship.

A 'material hermeneutical' or 'actor-network' view would represent a substantial shift away from the dualist and mentalist position of understanding order and intelligence as subsisting in either a mental or electronic informational realm. Here the power of knowledge and action would inhere in networks of organised human and nonhuman, or social and technical, factors. It would also require a substantial shift in our methods of mapping, away from the orthographic representation as a privileged objectivity, and towards the revealing of a structure of sociotechnical networks or 'systems' through which people order their knowledge of the world.

6 Putting 'World' Back into Relations

Dreyfus has proposed that reality is 'relative to us'. This 'relative to us' refers to the intentionality captured in the subject-object relation. This relation is no longer mediated between autonomous realms of mind and corporeality. Heelan has shown how the intentional relation becomes conditional on and mediated through the equipment we manipulate and through which results in the form of objects and knowledge are produced. The situations in which we know and do things are specific instances of this technically mediated intentional relation. This makes situations technically specific and multiple. They become pre-prepared in-the-world networks or spaces for acting and knowing, and artefacts in themselves. This testifies to the locally strategic nature of organised equipment and artefacts and to the spaces of knowledge, shared in communities they contribute to defining and maintaining. Not all of these will be high-tech though most of them probably were high-tech at some time in the past.

According to Heidegger, "equipment… always is in terms of its belonging to other equipment" (Heidegger 1962: 97). There is always a totality of objects and practices, a practical and equipmental whole, in which objects and practices define each other in their relations. This 'fitting' of object or subject into a context is called 'involvement' (Heidegger 1962). It is a fitting into an intelligible 'world' where 'world' refers not to the simple totality of physical objects, but to the nexus of functionality and intelligibility organised by our equipment on the one hand and practices on the other. Our environments are realised as an in-the-world relationality of meaningful objects. This is a view which emphasises the overwhelmingly artefactual nature of the world in which we know and do things (Mitcham 1995)[3] and then understands agency as distributed between agent and environment through the entrainment of the objects making up the environment with human activity (Clark 1997; Clark and Chalmers 1998; Hutchins 1995).

These situated collective understandings underpin the intrinsic sociality of the objects themselves that Latour insists on; the 'objective' world around us is a world of objects entrained with and in relation to us. Abstract distinctions between the subjective and the objective are not coherent in this view. Rather what is significant is the organised relations of objects with subjects in settings embedding knowledge and the possibilities of action. Although people connect with artefacts and equipment individually, the meanings and functions of things are commonly understood in settings and within the 'communities of practice' (Kuhn 1962) that use them. 'Worlds' of common understanding are social and normative, but it is through our individual emplacements in relation to things that our knowledge of them and their possibilities emerges. These situations are total 'worlds' that environ us because they are already prepared for involvement and action by being organised around subject-object relations. Involvement in them may also be restricted, however, to those who have the credentials and the skills to access, understand and use them.

Humans have managed to create multiple situations, multiple worlds, and multiple forms of life and the way they have done this is by forming multiple 'worlds' and spaces, technologically, within which objects, subjects and actions (things, knowledge and practices) are internally enabled and regulated.

On the basis of the technoscience perspective, an alternative relational phenomenology of place is possible: one in which emplaced objects constitute a material organisation through which our knowledge and action are mediated in situ. This alternative relational view would avoid any necessity of an 'interiority' of imagination; subjectivity is exteriorised and captured in the perfectly real relations between subjects and objects mediated through socio-technical networks. The significance of this is that we can now talk of a human world consisting of real things (and their spaces and places) that can be mapped and analysed. These objects participate in action chains and sequences—participation implying neither a subject-centred nor an object-centred perspective on action but one where the subject-object, actor-environment relation stands at the centre of the analysis.

---

[3] Carl Mitcham is only one of the more recent theorists who has argued that we live in a world of our own making and that the things we encounter in that world are artefacts. However, Mitcham also sees this as a something new whereas I am suggesting, along with others like Arendt and Leroi-Gourhan, that this technical character and capacity has defined the human condition throughout history.

7 Metageographies

The character of these sociotechnical 'worlds' may begin to be "conveyed in such expressions as 'the world of the parent', 'the runner's world', or 'the world of the treesitter'" (Thomson 2006: 447), but because our lives are emplaced, these 'worlds' will also be integrated with and constrained by structure inherent in territories and places. As we become urban citizens and neighbours, as we create national societies, and as we become a global multitude, the territorial dimensions of these communities and social identities will also be constructed. In fact all the 'worlds' we inhabit will be geographical in the sense that everything we do will have a situation and be emplaced. There is a structure to this geography, however, that relates to an historical formation and normative formalisation of the planet into what Lewis and Wigen have called 'metageographies'. These are "the set of spatial structures through which people order their knowledge of the world" (Lewis and Wigen 1997: ix).

As we can by now expect, these spatial structures will relate to objects entrained with the practices and ways of life of territories and places. These territories and places would be technically structured in such a way as to support the normative practices associated with them. The technics would mediate what the territories and places were and regulate and regularise the way they were enacted. This 'worldly' mediation would consist of an organised and strategic integration of technological, intentional and practice dimensions. The best way to understand this is by way of an example: we can get a sense of how these territorial spaces are formed and what they mean if we begin with Heidegger's carpenter's workbench and then extend our point of view outward. The workbench is a place constructed to emplace the tools and support the practice of carpentry. In this place there are a set of objects (the tools), a place or setting (the workbench), a skilled subject (the carpenter) and a practice (carpentry) which, when taken together, form a 'world' within which the carpenter's attention and intentions are captured, and within which he works and produces things. Tools and other objects take their meanings from this 'world', and a nail and hammer are entrained; they are to-hand and can be swept up in the movement of making a chair.

In this situation the equipment and space of the workbench disappear into the background—becoming to all intents and purposes not present to the carpenter—as his attention goes to the object of the action, the chair. Things are to-hand where their practical meanings and the actions they participate in coincide. At the same time the workbench can become an object of attention itself—and become present—when the carpenter is preparing it for working. Also, the nail, displaced by a few meters, will lose its entrainment and become present as an object that needs to picked up off the floor and replaced to where it belongs. In a human world to-handedness works next to presentness and relies on presentness to prepare and maintain places for action.[4] An object that is to-hand exists in what Heidegger calls a 'region'. Regions organise the things we are involved with in relation to practice. The region is a framing and referencing of objects, and a setting prepared for action. We see the region as we prepare or tidy up or otherwise get ready for action, but the character of the region is as a background to action and attention. We are capable of being involved in that 'world' as we are capable of stepping out of it in order to fix, prepare or tidy up.

But different settings and different 'worlds' also exist in relation to one another. 'Worlds' may be nested in relation to one another so that the nail is on the workbench; the workbench is in the workshop; the workshop is in the town (Heidegger 1962: 79). A nail may slip from being to-hand in one 'world' to being present in another as easily as slipping off the workbench—so there is no necessary, literal boundary between these 'worlds'. A nesting of 'worlds' or settings, is as organised and prepared in their relations to one another as each setting is individually. The nesting also allows us to quickly increase possible scales and scopes of action, taking us beyond the reach of our hands and our immediate lines of sight. The carpenter uses the setting of the street grid of the town in order to connect his workshop with the post office. The postal service is organised as a region of post offices such that a letter may pass from the post office in the street grid of one town to the post office in a street grid of another—and through that street grid to the letter's recipient. The postal service maintains and uses a system of pathways and roads to interconnect the towns in the county.

The workbench and tools, the workshop and its interior spaces, the urban street grid distributing the division of labour of an urban community, are each organisations of objects, subjects and practices that work together in stable and repeatable ways. Each 'region' constitutes a 'way of life' and a 'world'. Each gives a concrete measure of a scale of working associated with its own objects. But each region limits us to doing things in one framing and at one measure or scale, while different settings are connected to one another in ways that allow them to be negotiated sequentially in chains of actions that span multiple regions. These regions are articulated so we may move outwards, and 'vertically' through 'levels', beyond the reach of hands and vision. The whole 'structure of places' is a tight fit of territorial units of different scales matched with 'communities of practice' that know how to use them. All this facilitates a flexible cross-scalar coordination of social and economic activity, ultimately across vast distances. It is through this relatively low-tech (certainly in its original forms) topology, or structure of places, rather than through contemporary high-tech communication and mobility technologies that 'space' as distance is and has always been 'annihilated' (Harvey 1989).

---

[4] Heidegger used the terms *zuhanden* and *vorhanden* for these different conditions of objects as being respectively engaged with the subject as part of the action, and present to attention as an obstruction or in the course of preparation or maintenance of settings for action (Heidegger 1962: 135–148).

'Communities of practice' already know how to locate themselves and do things in the appropriate networks. Newcomers learn these things and join communities of rush-hour commuters or tourist shoppers or creative-region entrepreneurs. This is not by reference to a set of cartographic or geodetic coordinates but by a topological frame of referencing of 'objects' and places alongside, inside or outside specific settings. These alongsides, insides and outsides relate to the networks and not to territorial boundings. They are organised in networked metageographical 'levels' and scalar types. The nail is on the workbench in alongside relations with the hammer and chisel; the bench is in the workshop in alongside relations with the office and wood store. The workshop is in the town in an alongside relations to home, the post office and the inn the carpenter visits. Rather than dealing with universals of space and scale we are dealing with particular, located, scaled and delimited organisational networks, and the equipment and know-how to enable intentions and actions.

At the urban level the street grid is normally considered to be a medium for accessibility between objects already defined. What we see in the account I have just given is something different: the grid is the pre-condition for the definition of the objects concerned. In a 'space of flows', the nodes are ontologically transparent objects that define places and provide the datum; here the grid itself is the datum, specifying scale, producing the nodes and defining the elements of the town.[5] It is by being framed in the grid that the carpenter, baker and blacksmith are identified and defined as constituents of an urban 'world'—as urban parts of an urban whole. At the same time, other 'grids' at different 'levels' and scales identify tools as parts of a workbench, furniture and stores as parts of a workshop, and a network of towns as parts of a postal region. The network positively organises rather than passively connecting and its space is not a surface across which flows are traced, rather it is a topology and networked 'world' of meaningful relations that underwrites effective places and practices.

8 An Alternative Phenomenology of Places

Metageographical structures construct scale alongside place. So all places come integral with their scales as a direct corollary of the structure of places we understand collectively and objectively. We could start by imagining networks of small things and places, medium sized things and places, large and extra large things and places (Prytherch 2007). However, it's not the sizes of the things but the scopes of the wholes they are part of that defines the scales of metageographies. These scopes, and the metageographies themselves, are normative and historically

[5] What the 'flat ontologies' of Latour and others like Marston et al. (2005) miss with their networks freely connecting heterogeneous subjects and objects across scale levels, is the dependency of the objects (and subjects) themselves on scaled networks of objects (and subjects) for their definition as existent things.

constructed. They are cities, regions, nations, and centres, suburbs and neighbourhoods, but expressed not as objects in isolation but in networks of parts that make up wholes. We think of them as autonomous things that have always been there, but they are historically constructed and naturalised rather than natural. We use these both to access and to order our activities in networks and also to relate to other places through their part-whole relationships with other networks at other scales and scopes. The best studied examples of such networks are world-city networks (Friedmann 1986, 1995; Beaverstock et al. 1999). Here 'world' and 'city' are co-constitutive and interdependent, whole and part, mutually constructed in the network. 'World' is itself a metageographical construction comprising a network of cities and 'city' derives its meaning from its relations in the network.

These are not the only part-whole structures defining territorial entities; a smaller-scaled but formally analogous 'city-neighbourhood' network is a regular feature of European industrial cities. Here again, part and whole ('neighbourhood' and 'city') are mutually constructed in the network (see Read forthcoming). At smaller scales still, streets or houses, would be the parts making up the whole of the neighbourhood, while rooms are the parts making up the house. These structures have been imagined as bounded territories, and some have borders that are zealously defined and guarded, but all of them are constructed and given practical reality for their users in the networks through which they are enacted and used. Worlds, empires and trading leagues are historical constructions, formed in networks of cities; counties and principalities have appropriated and organised territories at smaller scales as networks of towns; nation states are a relatively recent invention and construction intended to stabilise shifting political authorities and allegiances, again constructed and organised around a network of urban centres; inner city neighbourhoods were constructed as parts of the industrial city, and out-of-town centres and suburbs were invented and constructed as parts of the post-industrial metropolitan city-region.

The neighbourhood is in the city and the city in the metropolitan region, nation and world and these places exist in 'vertically' structured nested relations with one another while neighbourhoods and cities relate 'horizontally' with other neighbourhoods and cities. Each one of these metageographies is a network in itself and each is part of a larger-scaled network and contains smaller-scaled networks. This is topologically a stacking of layers so that each of these metageographies exists in organised relations with others. But this is a 'layering' of different network spaces rather than of territorial units bounded in a universal space and in 'Russian doll' relations with each other. There are no necessary boundaries between the layers and one may slip out of the part and into the whole as easily as passing from a neighbourhood back street into a city main street. At the same time it is not possible to confuse the street with the neighbourhood or the neighbourhood with the city, or the city with the world for that matter, because the part-whole structure is at the same time a knowledge structure and a structure constructed and realised in the world. It is one of those 'concrete abstractions' that exist simultaneously in knowledge and reality and give the lie to the Enlightenment 'bifurcation' of nature.

An alternative phenomenology of places would deal with regions and places as sociotechnical networks which include emplaced objects, the technical 'grids' that connect them and the 'communities of practices' who know and use them. It would deal at the same time with the ways technical 'grids', networks of places, communities and practices are articulated with one another over scale differences. A different phenomenology of places would focus on places in networked 'regions' and the ways they support practice and object 'worlds' as sociotechnical systems. We live environed by such 'worlds' and systems. Karin Knorr-Cetina has in a research project which has spanned two decades investigated the networks and cultures of global financial trading. A 'world' is created in a high-tech construction involving optical cables and satellite links, along with a lot of expensive computing hardware and software, to present the objects of the global financial market real-time to traders wherever they have access to a terminal (Knorr-Cetina and Bruegger 2002). Financial traders do and see things in a technical system that goes far beyond technics to be acutely tuned to their individual perceptions and conceptions, and to a culture and practice of trading that spans the globe. Intention and meaning have been 'shifted' into equipment and into what Knorr-Cetina calls 'synthetic situations' (Knorr-Cetina 2009) in offices and on trading floors. There is a 'world' created in this sociotechnics that seems to 'annihilate' space and time. This 'annihilation' is, I have argued before (Read 2009), nothing more than a highly designed and maintained effect of the technics, and this 'world' doesn't exist in isolation; it is linked to other spaces so that the 'world' of global trading may be exited and the trader can use the office, or can exit the building to use the street network to find the metro and go home. Each of these 'steps' in a chain of activities is a transition from one situation, setting and 'world' into another. The space of global trading requires both access and security through portals with the spaces of offices, buildings, streets, public transportation systems, and so on, and each of these portals will be as technically and socially designed and effective as the spaces themselves.

The network of global trading is undoubtedly a powerful technology, but whether technologies like these imply anything about a society any more fundamentally 'new' than mercantile or industrial societies were is a matter for debate. Particular 'worlds', even contemporary global ones, may be less influential on their own than they are in their topological relations with other 'worlds'. They may define their power more in the ways access is regulated through portals with other networks than through their own particular technical and social operations and scopes. The contemporary distinction between 'spaces of flows' and 'spaces of places' falls away as global networks become articulated with other very everyday networks and places.

One of the functions of movement infrastructures like road, rail, metro and airline systems will be to define places and coordinate metageographies in their scales and scopes. For a London Underground user, London is the famous underground map realised. The infrastructure is invested with a diagramatic topology and realises a specific place-structure. This structure founds a 'world' through a particular technics and culture of travel by underground, which includes access, ticketing, information and security procedures. But it connects also to significant places that serve as portals to more global spaces, and responds to an overground network, a pre-existing structure of places to which the underground adds structure and emphasis.

It connects to the street grids of the neighbourhoods centred by underground stations, and it regulates profoundly the ways these different metageographies and cultures are articulated with each other. The equipment here may not be as high-tech as in Knorr-Cetina's example, but is no inconsiderable matter. Beside the transportation systems themselves and their signalling, scheduling and other support systems, there are public facilities, shops, offices and housing built in a systematic relation to transportation systems, business and industry to which employees, suppliers and clients need to be connected, and all manner of other technical and support systems, including street and line maintenance, energy, water and drainage systems, stairs and lifts, street and passenger signage that need to be incorporated. All of this is maintained in a high state of efficiency and order, reflecting the strategic role of this regulating and regularising infrastructure.

Cities, regions, nations, worlds and their parts are articulated, consolidated and stabilised in airline, underground, road, rail and other technical movement grids, but what is real is not just the material but also the coherence-giving, sense-making structures of part, whole and scale. The effect is that today the whole is locked into synoptics territorial users understand, with well-known places made visible and present locally and trans-locally. London is indivisibly and several times over, percept, concept, technique, object, network, culture and practice, a patch of integrated and strategic territorial order historically constructed and realised. The structures that make it up differ from and precede the cartographic map. They order the world according to a different order of precepts than the distances, geometries and boundings of the map. These constructions have their own remarkable levels of stability and persistence as place-structures, and as coordinated diversities of cultures and practices. Different individuals and different cultures and practices share, before all else, these structures of places as common knowledge. Taken together, these structures are a practical objectivity which informs every movement and every action—as well as every further intervention—in our human world.

This reassessment of phenomenology centres our attention on the historical construction of our anthroposphere, which includes a topological 'structure of places' organised by a layering of infrastructural 'grids' into a set of normative 'levels' which have a metageographical, intelligibility-giving and 'community of practice' defining, character. 'Structure of places', 'grids' and 'levels' are perfectly objective and mappable and are proposed as the foundation of a different phenomenological urban and geographical model. Further contributions will set out these structures in real examples in order to assess their potential as an urban and geographical model.

References

Agre PE (1997) Computation and human experience. Cambridge University press, Cambridge
Arendt H (1970) The human condition. Chicago University Press, Chicago
Boden D, Molotch H (1994) The compulsion to proximity. In: Friedland R, Boden D (eds) Nowhere: space, time and modernity. University of California Press, Berkeley
Beaverstock JV, Smith RG, Taylor PJ (1999) A roster of world cities. Cities 16(6):445–458
Brooks RA (1997) Intelligence without representation. In: Haugeland J (ed) Mind design II. MIT Press, Cambridge
Castells M (1989) The informational city: information technology, economic restructuring, and the urban regional process. Blackwell, Oxford
Castells M (2000a) The rise of the network society, the information age: economy, society and culture, vol I. Blackwell, Oxford
Castells M (2000b) End of millennium, the information age: economy, society and culture, vol III. Blackwell, Oxford
Castells M (2004) The power of identity, the information age: economy, society and culture, vol II. Blackwell, Oxford
Chapman A (2004) Technology as world building. Ethics Place Environ: J Philos Geogr 7(1–2):59–72
Clark A (1997) Being there. MIT Press, Cambridge
Clark A, Chalmers D (1998) The extended mind. Analysis 58:7–19
Dreyfus HL (1965) Alchemy and artificial intelligence. Report to the RAND Corp. http://www.rand.org/content/dam/rand/pubs/papers/2006/P3244.pdf
Dreyfus HL (2007) Why Heideggerian AI failed and how fixing it would require making it more Heideggerian. Philos Psychol 20(2):247–268
Durkheim E (1951) Suicide: a study in sociology. Free Press, New York
Edwards PN (2003) Infrastructure and modernity: force, time, and social organisation in the history of sociotechnical systems. In: Misa TJ, Brey P, Feenberg A (eds) Modernity and technology, pp 185–225
Friedman J (2007) Global systems, globalization, and anthropological theory. In: Rossi I (ed) Frontiers of globalization research: theoretical and methodological approaches. Springer, New York, pp 109–132
Friedmann J (1986) The world city hypothesis. Dev Change 17:69–83
Friedmann J (1995) Where we stand: a decade of world city research. In: Knox PL, Taylor PJ (eds) World cities in a world system. Cambridge University Press, Cambridge, pp 21–47
Fodor J (1981) The mind-body problem. Sci Amer 244:114–123
Hägerstrand T (1978) Survival and arena: on the life-history of individuals in relation to their geographical environment. In: Carlstein T, Parkes D, Thrift NJ (eds) Human activity and time geography. Timing space and spacing time, vol 2. Wiley, New York, pp 122–143
Harvey D (1989) The condition of postmodernity. Johns Hopkins University Press, Baltimore
Haugeland J (1997) What is mind design? In: Haugeland J (ed) Mind design II: philosophy, psychology, artificial intelligence. MIT Press, Cambridge
Heelan P (1977) Hermeneutics of experimental science in the context of the life-world. In: Ihde D, Zaner RM (eds) Interdisciplinary phenomenology. Martinus Nijhoff, The Hague
Heidegger M (1962) Being and time (Tran: Macquarrie J, Robinso E). Blackwell, Oxford
Hughes TP (1987) The evolution of large technological systems. In: Bijker W, Pinch T, Hughes T (eds) The social construction of technological systems: new directions in the sociology and history of technology. MIT Press, Cambridge, pp 51–82
Hutchins E (1995) Cognition in the wild. MIT Press, Cambridge
Ihde D (2009) Material hermeneutics: reversing the linguistic turn. Paper presented at the annual meeting of the 4S annual meeting—abstract and session submissions. Hyatt Regency Crystal City, Crystal City, VA. 2013-05-24 http://citation.allacademic.com/meta/p371011_index.html
Ihde D (2010) Phenomenologists and robots. Embodied technics. Automatic Press, New York

Knorr-Cetina K, Bruegger U (2002) Traders' engagement with markets: a postsocial relationship. Theory Cult Soc 19:5–6
Knorr-Cetina K (2009) The synthetic situation: interactionism for a global world. Symbolic Inter 32(1):61–87
Kuhn T (1962) The structure of scientific revolutions. University of Chicago Press, Chicago 2nd edn: 1970
Latour B (1992) Where are the missing masses? The sociology of a few mundane artifacts. In: Bijker WE, Law J (eds) Shaping technology/building society: studies in sociotechnical change. MIT Press, USA, pp 225–258
Latour B (2005) Reassembling the social: an introduction to actor-network-theory. Oxford University Press, Oxford
Leroi-Gourhan A (1993) Gesture and speech. MIT Press, Cambridge
Lewis MW, Wigen KE (1997) The myth of continents: a critique of metageography. University of California Press, Berkeley
Lynch K (1960) The image of the city. MIT Press, Cambridge
Marston S, Jones JP III, Woodward K (2005) Human geography without scale. Trans Inst Br Geogr 30:416–432
Mitcham C (1995) Notes toward a philosophy of meta-technology. Philosophy and Technology 1(1–2)
Moreno JL (1934) Who shall survive?. Nervous and mental disease publishing company, Washington DC
Pylyshyn Z (1980) Computation and cognition, issues in the foundations of cognitive science. Behav Brain Sci 3(1):111–169
Prytherch DL (2007) Urban geography with scale: rethinking scale via Wal-Mary's 'geography of big things'. Urban Geogr 28(5):456–482
Read SA (2008) Technicity and publicness: steps towards an urban space. Footprint 3:7–21
Read SA (2009) Another form: from the 'informational' to the 'infrastructural' city. Footprint 5:5–22
Read SA (2013) Technology and the body public. Inflexions 6 (www.inflexions.org), pp 188–211
Robertson R (1992) Globalization: social theory and global culture. Sage, London
Scott J (2000) Social network analysis: a handbook, 2nd edn. Sage Publications, London
Seamon D (2000) Phenomenology, place, environment, and architecture: a review of the literature. http://www.arch.ksu.edu/seamon/articles/2000_phenomenology_review.htm
Simmel G (1955) Conflict and the web of group affiliations (Tran: Wolff K). Free Press, Glencoe
Thomson I (2006) Environmental philosophy. In: Dreyfus HL, Wrathall MA (eds) A companion to phenomenology and existentialism. Blackwell, Oxford, pp 445–463
Wasserman S, Faust K (1997) Social network analysis: methods and applications. Cambridge University Press, Cambridge
Webber MM (1963) Order in Diversity: Community Without Propinquity. In: Wingo L Jr (ed) Cities and space: the future use of urban land. Johns Hopkins University Press, Baltimore, pp 23–54
Webber MM (1964) The urban place and the nonplace urban realm. In: Webber MM et al (eds) Explorations into urban structure. University of Pennsylvania Press, Philadelphia
Webber MM (1996) Tenacious Cities. In: Conference research notes: spatial technologies, geographical information and the city. Baltimore, MD
Wellman B (2001) Physical place and cyber place: the rise of networked individualism. Int J Urban Reg Res 25(2):227–252
Winograd T (1972) Understanding natural language. Academic Press, New York
Wright WDC, Stewart DH (eds) (1972) The exploding city. Edinburgh University Press, Edinburgh
Varela F (1996) Neurophenomenology. A methodological remedy for the hard problem. J Conscious Stud 3(4):330–349

The Cartographic Representation and Analysis of (Slovene) Writer-Careers: Methodology and First Results of the Slovene Project

Urška Perenič

Abstract This paper is part of the first phase of the research project "The Space of Slovenian Literary Culture" (2011–2014). One of the central aims of the project is to map, with the help of GIS, the literary biographical data of prominent Slovene writers from the beginnings of aesthetic production in the Slovene lands (1779) until 1940, before the WW II began in the Slovene lands. Data selection for mapping entries and the means of their representation follow critically the more recent achievements in literary geography (H. D. Schlosser), whose key work on literary mapping was published in 1983. The literary maps are extremely important to us from the point of view of the series of biographical data selected. Taking their solutions into account together with the purposes of the Slovene project, we offer some suggestions for the formation of thematic analytical maps to facilitate a contemporary areal analysis of Slovene literary culture.

1 Short Presentation of the Slovene Project: Entry Mask for Biographies

One of the three main goals of the project "The Space of Slovenian Literary Culture" (2011–2014) is to map, with the help of the geographic information system (GIS), relevant biographical data of Slovene writers from the beginnings of Slovene literary production (*Pisanice od lepeh umetnost* 1779) until 1940, before the WW II began in the Slovene lands. The database includes 323 biographies. Whereas comparable and earlier attempts at mapping literature focused almost exclusively on the production of literary communication or on the writers' lives, approximately only two-thirds of the biographies in the Slovene project are

U. Perenič (✉)
Department of Slovene, University of Ljubljana, Ljubljana, Slovenia
e-mail: urska.perenic@ff.uni-lj.si

S. Rau and E. Schönherr (eds.), *Mapping Spatial Relations, Their Perceptions and Dynamics*, Lecture Notes in Geoinformation and Cartography,
DOI: 10.1007/978-3-319-00993-3_9, © Springer International Publishing Switzerland 2014

concerned with writers, while one-third will focus on translators, literary critics, editors, publishers, printers, librarians, and literary scholars who published their main works or otherwise earned a place in the literary field before the year 1940. These people were in charge of publishing, distributing, printing, reviewing, preserving, and processing literature.[1] The various stations in their life paths are to be found in different locations (birthplaces, places where their works have been published, etc.). Space has for a very long time been of interest to various disciplines (from physics to the contemporary humanities); in the present context, which links literary studies and geography, space is understood primarily as a complex of natural, physical, political, economic, administrative, and demographic structures. We are not dealing with a deterministic understanding of geographic space. We assume that the geographic space is one of the chief factors in literary culture's dynamics.[2] (see Perenič 2012b: 265) Literary zones, margins, centers etc., are the spatial aspects (e.g., literary margins are less productive than literary centers which refers to the places where literature was written, published etc., and the term *literary networks/circles* refers to the connections between different literary actors). The goal of the project is geospatially to analyze the extent and density of literary networks and writers' circles and groups, from which we can see where the most productive literary zones were located within their ethnic territory and beyond; also to investigate what were the literary margins and centers and which locations produced, absorbed, or hindered literary potentials and energies. The added value of the project is to use maps as a cognitive tool—the project may provide new (unexpected) insights into Slovene literary life (e.g., the visualization of the places of literary vocation has already drawn attention to the importance of some places that were not until now recognized as Slovene literary centers or zones).

For the purposes of mapping Slovene literary culture, a mask has been designed for entering statistically relevant literary and historical data pertaining to writers' lives. Entering the biographies with stress on the spatial patterns in writers' lives represents the pilot phase of the project, which has, for the most part, already been part completed and should show which literary-historical data are (the most) relevant for entering into thematic maps. The structure of an entry for the biographies has anticipated the following literary and historical data:

- The identification number of the author
- Last name
- First name
- Additional names (pseudonym and the like)

[1] The term *literature* is thus understood from the viewpoint of modern systems and contextual methods. It is conceived of as a partly autonomous system in which texts are inseparably connected with the literary activities of production, distribution, reception and processing, the media and institutions (Schmidt 1980; Perenič 2008, 2010).

[2] It is also impossible to overlook the reverse influences on (social) geographic space; how literature has through symbolic representations influenced the apprehension of the ethnically Slovene space.

- Main activity: Poet, narrator, dramatist, writer of semi-literary genres [essay, (auto) biography, travelogues, diaries, memoirs], juvenilia, critic, translator, editor, publisher, printer, librarian, literary scholar
- Side activity: Poet, narrator, dramatist, writer of semi-literary genres, juvenilia, critic, translator, editor, publisher, printer, librarian, literary scholar
- Date of birth
- Date of death
- **Place of birth**[3] (maternity hospital, the place of mother's residence at the time of birth)
- **Place of death** (actual place of death)
- **Location** of the grave
- Sex
- Social origin (social class of the parents or the father)
- Ethnic origin (mother's and father's)
- Mother tongue
- Type and name of the secondary school
- Place of secondary schooling institutes of higher education/university; field of study or the faculty
- Place of higher/university education
- Type and degree of education
- Vocation(s)
- Place of vocation (by years)
- Social affiliation after the completion of studies
- Political functions or activities
- Personal connections in the literary field
- Personal connections in other fields[4]
- Languages in which books were published
- Year and place of publication of the first edition
- Year and place of publication of the last book (during the author's lifetime)
- Place where most of books were published
- Place of residence at the time of the first publication
- Place of residence at the time of the publication of the last book
- Magazines in which the author was published
- Publishing houses where the author was published
- Memorial events (memorial day, holiday, memorial march), literary societies, literary prizes

[3] Highlights refer to the so-called spatial data that will be visualized.

[4] Connections in the literary field refer to personal contacts, circles of friends, to teaching and mentor relationships. Non-literary connections refer to authors' contacts with important people who were active in other fields (other areas of artistic endeavour, science, religion, politics, etc.).

2 Pioneering Achievements in Literary Geography with Special Emphasis on the Methodology of Horst D. Schlosser's (1983) *dtv-Atlas zur deutschen Literatur: Tafeln und Texte*

The use of maps in literary studies has a relatively long tradition, extending from the beginning of and throughout the 20th century. Our attention will be directed to the rather narrow scientific field of literary geography and not so much to the popular contributions in this field, even though, when mapping literary culture, it might be possible to consult tourist guides, literary works equipped with maps (either by the writers or by the publishers),[5] or in literary atlases designed for a wider readership.

There are two methodological orientations in literary geography. The first results from the biographies of the authors and is focused on their birth places and places of residence, which are then shown on thematic maps. The second methodological orientation which currently prevails but is just a parallel goal of the Slovene project is primarily interested in two areas: firstly, in spaces represented in literature and secondly in their connections with the relevant geographical areas.[6] The first orientation is older. That can be on the one hand associated with the prevailing positivist orientation at the end of the 19th century, when an important role was ascribed to the circumstantial factors of literature (e.g., the effect of the local environment on the author). On the other hand, we need to know that we are also dealing with a "blood and soil" (*Blut und Boden*) ideology that emerged from the racism and nationalism of the late 19th century. This orientation mainly covers the pioneering achievements in literary geography of such scholars as Nagel (1907), Nadler (1912–1927). Taking into account the ideological background of these achievements and the fact that their concept has already been discussed extensively in another paper of mine, we will discuss biographical data and the possibilities for mapping in connection with another core theme and with more recent work in the field of the literary geography.

Horst D. Schlosser's (1983) *dtv-Atlas zur deutschen Literatur* is a standard work for mapping literary-historical data and concepts. In comparison to the previously mentioned books, this atlas provides different (non-deterministic) perspectives on spatial factors and encodes more types of spatially linked biographical data, all of which is of the greatest interest for the Slovene project from the point of view of the series of biographical data selected and for the formation of thematic analytical maps. The book traces the history of German literature from its beginnings up to the present—by century and by individual authors together

[5] To name one representative example—the legendary Defoe's robinsonade with its pictorial map.

[6] Miran Hladnik and Jerneja Fridl: Space and its geographical presentation in Slovene historical narratives (2012).

with their works, and—also by literary tendencies, types, and genres. The graphical representations comprising 116 items do not only include maps but also numerous color graphics, schematic displays, and temporal diagrams. The maps appear on the first two pages of the atlas (Schlosser 1987: 16): one refers to the pre-literary eras and shows the spatial arrangements of German dialects and speech types from the 8th to 10th centuries; the second is marked with the locations where runes had been discovered. The edition contains thirty-one maps altogether, which represents approximately 27 % or a little more than a quarter of all the pictorial displays of the development of German literature. The literary maps fall short of the standard expectations of maps because they are not equipped with physico-geographical elements such as mountains and waterways, or with socio-geographical elements. Political or administrative territorial units are unclear or not indicated; only the continental and sea surfaces can be discerned in outline, but without great difficulty. The indicators for some larger locales provide some orientation. From the Slovene project's point of view, it is interesting to cast a glance at the data that has been encoded and mapped. In this regard, the maps could be divided into two large groups. (a) One type of map is of a general nature as it contains and displays data on the political or administrative divisions of a particular territory, social, religious, and other kinds of conditions for different historical eras. Since we are dealing with a literary atlas, we would expect these representations to be more frequently combined with literary-historical data in such a way that it would be possible to use the map for analyzing literature in a given set of political or administrative, physical, natural, economic, and social circumstances.[7] (b) The second type of map relates to the literary-historical data. The data selected and mapped are of different types, so that all the maps are not equally interesting for the Slovene project.

- One group consists of maps thematizing the pre-literary eras and showing the oldest written traditions, citing and placing written sources from the 8th to 12th centuries. Sometimes they are combined with the aforementioned general geographical maps—e.g., the Early Medieval tradition is depicted from the perspective of the monastic reforms.
- The second group is represented by thematic maps that cover the Middle Ages, but the focus is on the literary types and genres. A typical example would be the spatial representation of troubadour poetry. A special map shows the areas of troubadour poetry pertaining to the younger generation. Here we could include maps to show the locations of folk literature (e.g., the locations of the oldest narratives) and those that spatially define the mystical texts.
- The third group of literary maps presents cultural, artistic and literary movements. They are the thematic maps showing humanism in Germany, centers of

[7] The atlas also includes, more radical changes which have been separately mapped into socio-political relationships connected to revolution and other forms of progressive social revolt that influenced the culture and literary events. For example, maps showing the political division of the empire, the social and religious revolts, or the reactions to the tumult of the French Revolution.

Baroque literature, centers of the German Enlightenment, Storm and Stress, Romanticism, literary circles of Expressionism, etc.[8]

- The fourth and final group consists of maps of the biographies of the more important German authors.

The latter are of most interest to us from the point of view of the series of biographical data selected and because of the forms and techniques of representation. The first case of spatial representation of a life and work is that of the Medieval German poet Walther von der Vogelweide (Schlosser 1987: 62). The map has a time line on the left, which represents key events in the author's life; on the right side, there are explanations of the three types of symbols that are used. The yellow circles with a question mark in the middle signify twenty possible places of the author's birth; pink circles with an image resembling a castle tower represent important visits to European courts (including one in Vienna); and an arrow with a thin red line on its tip shows the extent of his travels, so that the reader can picture the territory in which the author moved during his lifetime. Unlike the arrow denoting the direction of movement in space, the other two symbols (even the circle with the iconic image of the castle) are arbitrary and unconventional. This is also evident by the fact that we will not find them on the maps in the following text. With regard to the biographical data selected for representation, it is to be expected that the (supposed) birthplace should be shown on the map and that the movement of the author in space should be thematized, since these aspects ought to be important for the genesis of his literary works. The Slovene project is similar, but we follow the author's movement through different stations, which would be impossible for the life of the famous German poet (e.g., publishing locations).

The next map thematizes the biography of the late-Medieval German poet Oswald von Wolkenstein (Schlosser 1987: 82). The emphasis is on the destinations of his journeys in the 14th and 15th centuries, which supposedly influenced the development of the poet's talent. In comparison to the latter map, only one type of data is now prominent and the destinations mentioned are marked with pink circles. Here we can find some parallels with the Slovene project, which also envisages mapping data from individual authors' lives—e.g., a thematic map with all the locations where books in Slovene have been published throughout Slovene history.

More suggestions for the selection and representation of different biographical data are to be found in the depiction of Heinrich Heine's life (Schlosser 1987: 190). The map shows his mediation between French and German culture, including his connection between Paris and some German cities (e.g., Hamburg), where Heine was active either for a long or a shorter period of time. The biographical data selected thus refers to the places and types of employment. When

[8] In this group we could place maps of German post-war emigration, because we are similarly dealing with a larger group of people, only that here, in comparison to the artistic movements, people are connected by migration.

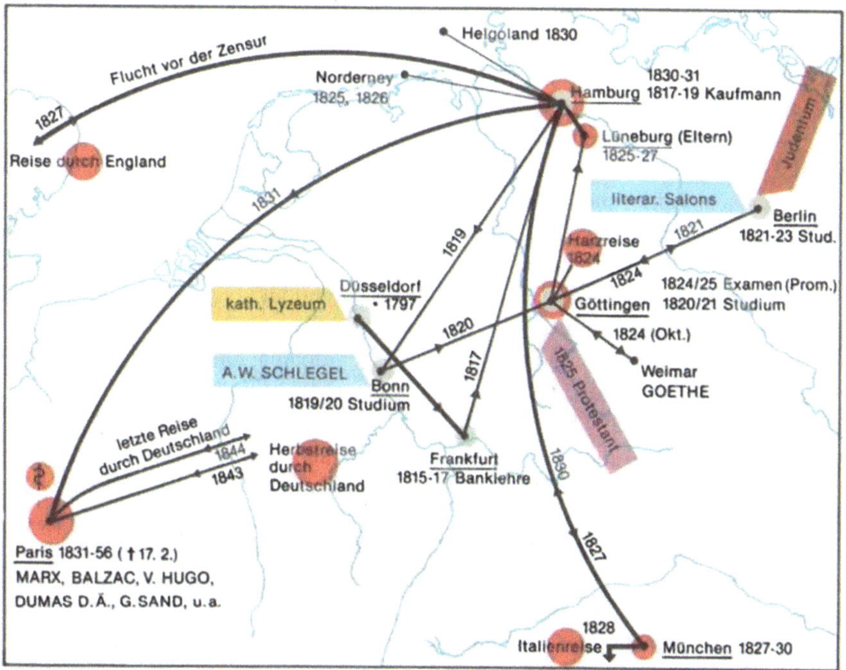

HEINRICH HEINE

Fig. 1 Heinrich Heine's life path (Schlosser 1987: 190)

explaining literary development and the spatial spread of literature, there is special attention given to occupation and place of employment in the structure of the entry for the Slovene authors (Fig. 1).

To sum up, three types of biographical data have been mapped—namely, movement or mobility in a particular space (study, employment, travelling), and the starting and the ending points in the authors' lives (birth [*] and death [†]).[9] In comparison with Schlosser's atlas, where only the locations of higher and university education are mapped, the Slovene project additionally maps the places of secondary education. On the one hand this makes sense, since universities were not very common in past eras and also not equally accessible to everybody; on the other hand, we know that many Slovene authors received a higher or university education outside their ethnic area. What the Slovene mask did not anticipate are journeys, which must be taken into account when explaining creative development. However, these kinds of spatial movements are implicitly included in some other rubrics, as in higher or university education (for which authors almost out of necessity and for different periods of time left their homes, at least until 1919) and

[9] The structure of the entry is more differentiated than Nagel's and Nadler's.

in the rubric "places of employment and work practice" (that led them to different locations and was a constituent part of their life and literary experience).

The other spatial models of biographies that can be of use when discerning the data series of spatially relevant biographical data are also included in Schlosser's diagrammatic depictions[10] of German literature. The first such depiction shows the life stations of Gotthold E. Lessing, where the emphasis is on the connection of literary-critical works and locations important for their formation (Schlosser 1987: 142). In such a way, the diagrams, much like maps, generate an idea of the author's real life space. In addition, they attempt to convey the amount of time the author spent at a certain location, what he was doing and creating there, and with whom he established connections. All of this is mostly given in the form of inscriptions. This might suggest which other biographical data must be considered when explaining literature. With Lessing, an interest in literary vocations (e.g., a dramaturge in Hamburg 1767–1770; a librarian in Wolfenbüttel 1770) is demonstrated for the first time; they are a constituent part of the Slovene project's entry masks and are, for example, marked with vivid green circles. The diagrams also show literary connections, but they are, a little surprisingly, the same as the places of education and are marked with pale green circles. If we take specific examples from Lessing, Berlin and Leipzig, where he studied, then they are marked with the same circle as, for example, for Voltaire, whom he met in 1750. That confirms the assumption that the sign system is not uniform because the same symbol refers to two different sets of data.[11] Next, the titles of literary works ascribed to particular places are marked in vivid green; but instead of the circles, rectangles are used. A lightning symbol, representing tension in cultural-literary relationships (e.g., Lessing-Gottsched) appears three times in the display. Since the literary relationship was previously marked with a circle, it would be better to remain with the same symbol, since we are dealing with a subtype of a personal connection. In the Slovene mask, both cultural and literary relationships are also highlighted but there is a common field established for the entry, which is meant to depict networking.

Goethe's life is depicted according to the same principles (Schlosser 1987: 158). Numbered locales containing data about studies and occupations are shown; the titles of his works[12] are given, and information is supplied concerning his acquaintances and circles of friends as well as spiritual and artistic

[10] In all, there are thirteen examples and they discuss the (dependent) relationships among the various variables; e.g., the dynamics of the literary types and genres in the passage of time or in the course of the literary phenomena within their temporal and spatial coordinates.

[11] Therefore, misunderstandings are possible. From the enclosed text accompanying the representation, it is otherwise clear that Schlosser tried to show the critical conflict between the authors, even though the light green circle used for the place of study would rather suggest Lessing's study of the French Enlightenment writer's works.

[12] Similarly the relationships among the locations and the works of Jean Paul are shown (Schlosser 1987: 172). For instance, Schwarzenbach, where the writer was managing a private school, is provided with titles, since the teachers are the literary characters. This explains the role of a relevant place for literary planning.

initiatives.[13] Since the latter is also included in cultural relationships, there is no need additionally to burden the depiction. It could be indicated in the same way as above—that is, with a circle. The depiction of Schiller's movements is not very different; both geometrical symbols are used (a circle, a rectangle), only in different colors. Places such as Tübingen, Maulbronn, and Denkendorf are provided with information about studies, employment, and cultural-ideological orientations, which were important for the poet's artistic development.[14] Schiller appears as an influence on Hölderlin's poetry (Schlosser 1987: 170), since he published a fragment of *Hyperion* in *Thalia*. His literary contacts included Fichte and the publisher Cotta, both of whom influenced the rewriting of Hyperion and both showed an interest in publishing the work. For the first time attention is drawn to another kind of literary actor, in this case a publisher, which is a factor that receives greater attention in the Slovene project. Besides new types of data for mapping, next to the reference to Bordeaux we notice a new symbol—namely, the Rod of Asclepius—around which a snake is wrapped ((⚕)). This indicates a poet's mental disorder or a disease. The Slovene project does not have any special rubrics for the author's mental state. The information is, in my opinion, too detailed for the purposes of a map, and the spatial fixing would also be too demanding. A visual representation simply cannot include all possible data; it is necessary to select information based on what we actually want to present. Of further concern is the fact that, the Rod of Asclepius is a symbol of the medical profession.

Among the symbols, the musical symbol for a *quaver* stands out ♪; it appears with the German Romantic, E. T. A. Hoffmann (Schlosser 1987: 184). We can find it, for example, next to Posen, Warsaw, and Bamberg—places in which he set various works to music. The Slovene mask similarly anticipates entering and spatially arranging the authors' main and peripheral activities.

With regard to the tradition of mapping biographies, the fact that there are almost no connections among conscription series of spatially linked biographical data is more problematic than the specific manners of representation and defining the data for display. In other words, an evaluation of the correlational connection between them is missing. Instead, we are dealing only with some sort of

[13] Leipzig has data attached about legal studies; a timeline and the information about its rococo characteristics are also given. Strasbourg has indicated contacts with Herder; Wetzlar is supplied with information about the lawyer's clerk practice. From there an orange arrow points to *The Sorrows of Young Werther*, because the author supposedly received his first inspiration for this epistolary novel in Leipzig. The number 6 indicates Goethe's trip to Switzerland, which took place shortly before his departure for Weimar.

[14] The information about his studies of law and medicine is attached to Stuttgart, the performance of *The Robbers* on a stage in Mannheim, where we also find the information about his essays or speeches about drama and theater art (*Die Schaubühne als moral. Anstalt*). It is evident from the representation that in Dresden, which he visited on account of his mother's unbearable material situation, he struck up friendships. We can also see how many and which works were written there. If we follow the arrow, we arrive at Weimar, where he established more intimate contacts with Herder, Wieland, and others.

conversions of data from biographies into pictorial representations. They are supplied with all sorts of information[15] about the authors' lives (locations of activities, locations important for various influences, locations of connections, etc.), and it is unclear whether these factors actually operate independently or interdependently, and thus they do not explain the spatial dynamics of literature and its arrangement in a space, but are rather illustrations of condensed textual biographies. The entry mask of the Slovene project anticipates most of the data enumerated and it proposes some additional kinds, so that in light of the earlier projects, it has become a unique summation of both sets of data. Given the capabilities of modern GIS technologies, which enable the encoding of huge amounts of different (non)spatial data together with their representation, arrangement, and connection, literary scholarship should also, for the purposes of a quality and comprehensive solution to the problem, model the data in view of process models of literary communication,[16] and it should, in addition, anticipate relevant combinations of data as well as the correlations between them.

3 The Space of Slovenian Literary Culture (2011–2014): Proposals for the Compilation of Thematic Analytical Literary Maps

The thematic literary maps designed as an analytical tool for the spatial analysis of literature will focus on individual objects and will display the following spatially linked components of biographies:

- The network of birthplaces, prepared from the database of biographies. A synthetic (not the individual) map is intended, where information would be arranged as points, enabling a (new) apprehension of the spatial (local, regional, provincial) connection or "the origin" of the more important members of the Slovene literary culture together with their arrangement and density in space (through time).
  Maps might display particular (literary) historical eras (e.g., the creators of Realism by places of birth). However, for comparative purposes, when we are looking at multiple eras, the bar or Gantt map would be more suitable. They can show a number of authors for the era selected by locale, region, or province.
- The network of the places of death, also prepared from the database, might lead to a (new) apprehension of the spatial arrangement of the creative forces at the

[15] Consequently, a number of different characters are used—from ordinary arrows to musical symbols—which is not necessarily a factor to be deplored, but the problem lies with the standardization and with the functionality of displays.

[16] E.g., a connection between the place of publication and the literary activity or between the place of work and the memorial events or the connection of places of networking, writing and publishing etc.

end points of the authors' lives and, as it were, the gravitational centers of the national literature. A consideration of some other types of spatial biographical data, besides the aforementioned places of birth, would be the basis for assessing the movement or the variety of ways in which literature had spread. In testing different combinations of spatial data for the spatial analysis of literature, this combination is not likely to be an optimal one. Many other combinations of geospatial data need to be considered (see footnote 19 below.)

- The network of secondary school locations, together with educational institution locations, belongs to the wider context of literary culture. Analyzing the development of secondary education is interesting because the secondary school was a place to become acquainted with literature and also provided (in)direct incentives for literary production. A common map showing the locations of educational institutions prominent writers attended (for this purpose a point depiction is most suitable) could offer insights into the spatial arrangement of potential incentives or stimuli.[17] It would be possible to ascertain which schools or areas "produced" prominent authors and where (besides the authors' homes, places of employment, etc.) the other potential "birth places" of the Slovene literary culture are located.
- As to the co-creation of conditions for the development of literary culture and its spatial spread, the same is true for the network of higher or university educational institutions, even though the forms of education should be considered together.
- This includes, for instance, consideration of the connection between secondary schools and universities (*Hochschule* = institute of higher education; meaning unclear with the term 'higher' schools) where Slovene authors were educated. However, the so-called synthetic maps with point indicators are not the most suitable tools for considering the connections between schools (e.g., the connection between grammar schools, secondary modern schools, or technical schools and some types of higher education). We could use linear indicators, which again are not the preferred form for representing such a large number of authors. Another possibility is a schematic model, which shows the interrelations between the individual schools. In such cases there are also the individual maps that might show the educational paths of individuals or group members in a certain era; however, this does not conform to the tendencies of a corpus-based analysis of the data (from 323 biographies).
- Spatial components of biographies include areas of career paths that must be considered when we are assessing or verifying the modes of mobility of the literati and the spatial extent of literary culture. However, mapping careers, which is often connected with many different locations, actually belongs to the depiction of individual authors' life paths or of groups of creative writers. That does not mean it is impossible to show the work locations by using points on a map and taking into account all the biographical units of the corpus, but the

[17] Secondary education is not necessarily a (solely) positive factor.

research question in this case is different. A common map can offer primarily answers to the question as to which locales, regions, and provinces have through history created and attracted literary impetuses.

- Schematic or diagrammatic depictions (e.g., paths between locations and people referenced geographically are marked with arrows) are more suitable for the spatial information on the density of (non)literary networks. To show the eras and ways in which, for example, literary groups were formed, it is still possible to use the classical map (and the point display); however, emphasis is again not on the individual figure, but on the connections between the locations where a literary culture was developing.
- Similar dilemmas occur with (print) media and publishing houses (publishing network, societies),[18] which facilitated and helped to consolidate literary activity; they also belong to spatial extensions of biographies. In the context of the spatial development of literary culture, it makes sense to think about the thematic maps when we are dealing with the locations of media, institutions, and establishments; but there is still not enough data collected for them (the entries for media and institutions are for later phases of the project). Media and institutions can for the time being be discussed in connection with individual authors or individual life paths. Yet if we are interested in their numbers in connection with particular locations and areas (as with some other spatial data) and displaying media by era, then a graphic display is to be preferred. Visual depiction of media and institutions is, among other things, important for recognizing the degree of development of the (literary) communications' infrastructure with regard to its spatial and temporal dimensions.
- The spatial analysis of Slovene literary culture anticipates a visual depiction of individual literary memorial events—for example, literary pilgrimages. One of the goals is to determine how these events intersect the ethnic space and beyond. Cartographic practice teaches that a display of all the literary pilgrimages at the Slovene level is not possible or it might be insufficiently accurate; we would need a map with a bigger scale for such data.[19] Therefore, only a selection of journeys will be mapped—in connection with individual authors. Another aspect worth looking into is the connection between the spatial distribution of memorial events and other variables such as locations of death, education, and career. In this regard, a question arises as to whether it would be possible to include some kinds of memorial zones in which individual artists and their works are represented and which are, in fact, the elements of the cultural collective identity of a particular place, region, zone, and province.

[18] I have analysed the reading societies network from the standpoint of socio-geographic factors (demographic structure, administrative, political, judicial organizations, development of the educational system) in the 1860s (Perenič 2012a).

[19] For the constructive scientific debate about the possibilities and potentials of the literary mapping, I am much obliged to my colleague Jerneja Fridl, who is both a geodesist and a cartographer.

4 First Results of the Slovene Project

Below are the first examples of thematic maps that have resulted both from the pilot and second phases of research. First, we mapped thirty biographies, then an additional fifty-nine resulting in a total of eighty-nine, or one-third of all the biographies, according to birthplace and place of death. We did not select the authors, but included them after the completion of the editing process. In the pilot phase, it was still possible, to a certain extent, to conjecture how the features of the spatial dynamic of the development of Slovene literary culture would appear—that is to say, which places, regions, or former imperial provinces exported (at least at the endpoints of authors' lives) or imported creative energy. However, this was not possible to verify for three reasons, the first of which was the relatively small number of authors included. Second, it proved necessary to take into account the time frame because the role of individual parts of the territory, with birth and death dates as functions, had been changing throughout history. Third, other spatio-biographical data must first be mapped, and on this basis it will eventually be possible to derive the features of the spatial dynamic. The conclusions of the first segment of the project[20] can be summarized in two points. Regarding the birthplace, it was possible, though based on an admittedly small sample, to discern the role of Carniola (including Dolenjska, Gorenjska, and Notranjska), which has throughout its history been the most Slovene of the cities. As the relative prominence of historical Slovene provinces and regions varied through history, this factor needs to be reviewed against a larger sample and needs to take into account the different periods of Slovene literary history, during which the regions' roles changed. The map of death places permitted us to surmise a centripetal force in Slovene literary culture; the authors included often made their careers or ended them in the Carniolan center of the ethnic space. Because not all of the biographies have yet been edited, the hypotheses cannot be confirmed or rejected. We will attempt to test them on a larger sample, and to add certain other possibilities for which it would be more logical to deal with when based on the full complement of material

On further examination during the first phase of the research project, it can be seen that the relation between the regions and birthplaces favors Carniola, with a total of twenty-two authors, followed by Primorska, Styria, and Carinthia, with five (one of whom was born in Trieste), then two, and one, respectively.[21] The situation is similar with places of death. Carniola holds the top position, and it must be noted that almost two-thirds of the authors (i.e., nineteen) died in Ljubljana whereas only four were born there. The supposition of a centripetal force in Slovene literary culture is connected with this fact. Three authors died in

---

[20] I presented them at a comparative literature colloquium in Lipica (2012).

[21] Within Carniola, Gorenjska has nine authors, followed by Dolenjska and Notranjska with four. V. Vodnik, Z. Kveder, J. Cigler, and J. Murn were born in Ljubljana; S. Grum was born in Zasavje.

Primorska (again, one in Trieste), which almost matches the number born there. Urban Jarnik was born and died in Carinthia [in Potok in the Ziljska Valley and in Možberk on Vrbsko Lake (Wörthersee)], which suggested low mobility for Carinthian writers. Physical, economic, transportation, and other geographic factors would be relevant to check. Zofka Kveder, who was born in Ljubljana, ended her life outside her ethnic territory, in Zagreb, making her lifespan different from that of the majority of authors under study. From the perspective of Ljubljana, she was an outlier.

In the second research phase, which considered eighty-nine authors, once again Gorenjska (twenty-one), Notranjska (thirteen), and Dolenjska (ten) were in the forefront, though in a slightly different order, followed by central Slovenia (ten), thus making a total of fifty-four authors. Primorska followed, with twelve authors,[22] then Styria, with eight (Fan[n]y Haus[s]mann was born in Upper Austria). Five of the authors selected were from Carinthia, three were born in Bosnia-Hercegovina (O. F. Babler), Serbia (Iz. Cankar), and in County Istria in Croatia (A. Cerkvenik). In comparison with the pilot phase map, Prekmurje is better represented, with five authors, four of whom were Roman Catholics and there was one Evangelical clergyman. If we consider places of death, then the range of places is less diverse. However, eighteen of the authors died abroad—four in the U.S., and one in Argentina, Australia, Austrian Graz, Bosnia-Hercegovina, and Prague respectively; two died in Italy[23] and Zagreb; and four in Hungary. Ivo Grohar, from Dolnja Bitnja near Ilirska Bistrica died in the Dachau concentration camp. In this case, the place of death is by no means connected with literary activities, which once more highlights the fact that spatial data cannot be treated in isolation, but will have to be linked with non-spatial data (e.g., reasons for involuntary departure to Germany).[24] All of this must be born in mind in the case of Ljubljana and its environs, where forty authors, or almost half, died.[25] On the one hand, then, it is the last station in writers' literary careers; on the other hand, we must keep in mind other factors that might prevent Ljubljana, in any given case, from being seen as the literary center. It is quite telling, though, that markedly fewer authors died in Gorenjska (five),[26] Dolenjska (three), and Notranjska (four) than were born there, which indicates gravitation towards Ljubljana. The same would be true of Styria, Primorska (with the exception of Gorica), and Carinthia. From the perspective of mobility, Carinthians were the most loyal to their land, which partially

---

[22] Three were from Trieste, which at the end of the nineteenth century was becoming an important Slovene literary center (the women's newspaper *Slovenka* was published there starting in 1897). One author was from Štandrež in Goriško, which today is in Italy.

[23] Because we respect the time frame, we do not count F. Cegnar and J. V. Koseski, who died in Trieste. In contrast, J. Bagarič is counted among those who died in Hungary because Prekmurje was attached to Slovenia only several months later.

[24] The same would be true of F. Balantič, who died in battle at Grahovo in Notranjska.

[25] J. Burgar died in Šmartno near Litija.

[26] I. Hribovšek's place of death is unknown.

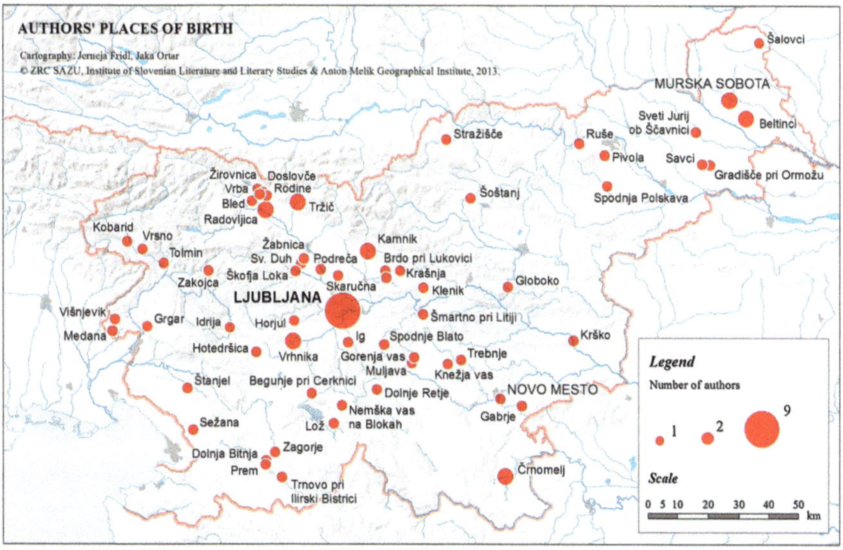

Fig. 2 Authors' places of birth. In the *bottom right corner* of the map, the Slovene authors born in Ljubljana, have been selected

Fig. 3 Authors' places of birth beyond the boundaries of today's state

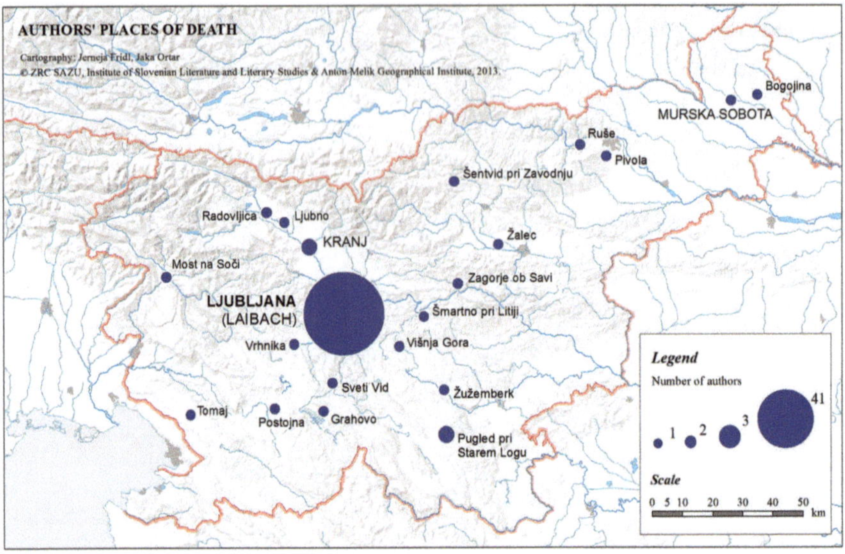

Fig. 4 Authors' places of death. In the *bottom right corner* of the map, the Slovene authors who died in Ljubljana, have been selected

Fig. 5 Authors' places of death beyond the boundaries of today's state

confirms the thesis we posited about the lesser mobility of Carinthian authors (Figs. 2, 3, 4, 5).

Since the aim of this research is more precisely to determine how literary actors moved between their birthplaces and places of death, the project will have to deal more with combinations of the two relevant factors rather than treating them separately. Non-spatial factors will also have to be taken into account. As noted above, the place of death will have to be combined with the motive for departure to a specific place, the last place of employment, or the penultimate stop in life. We imagine that in this way, it will also be possible to arrive at some spatial patterns in the historical development of Slovene literary culture (e.g., whether, after schooling, a career path tends towards home or away from it,[27] what the various employment paths were like, how life stages are linked to places of publication, and so forth). In this way, it will be possible to answer the question posed at the outset as to what were the most productive areas in terms of literature, and which locations or regions either engendered or received relatively more creative energy. We will also be able to answer the question as to what were the features of the spatial spans (e.g., the direction of the life path taken from birthplaces, and how places of schooling and employment, or employment and publication are linked), and/or whether it is possible in analyzing the space of Slovene literary culture to identify a common spatial developmental pattern or at least to recognize the predominant patterns of authors' movements. Therefore, it is necessary once again to continue to prepare maps together with the spatial analysis of literature, from which we anticipate more comprehensive results.

References

Nadler J (1912–1927) Literaturgeschichte der deutschen Stämme und Landschaften. J. Habbel, Regensburg

Nagel SR (1907) Deutscher Literaturatlas Die geographische und politische Verteilung der deutschen Dichtung in ihrer Entwicklung nebst einem Anhang von Lebenskarten der bedeutendsten Dichter auf 15 Haupt- und 30 Nebenkarten. C. Fromme, Dunaj, Leipzig

Perenič U (2008) Perspektive empirične sistemske teorije z vidika mlajše generacije—doslednost, odprtost, zanesljivost (Empirical systems theory perspectives from the viewpoint of younger scholars: consistency, openness, and reliability). Primerjalna književnost 31(2):113–135

Perenič U (2010) Empirično-sistemsko raziskovanje literature: Konceptualne podlage, teoretski modeli in uporabni primeri (Empirical systems research of literature: conceptual framework, theoretical models, and applicable examples). Zveza društev Slavistično društvo Slovenije, Ljubljana (Slavistična knjižnica; 16)

[27] As concerns places of employment, it will also be necessary to take into consideration non-spatial data (e.g., occupational profile) that can be pinpointed geographically. In the cases of priests, the Roman Catholic Church played a significant role in their settlement, sending them to serve in different places.

Perenič U (2012a) The reading societies network and socio-geographic dynamics. In: Perenič U (ed) Slavistična revija: Prostor v literaturi in literatura v prostoru/Space in literature and literature in space, 60(3). pp 383–400

Perenič U (2012b) Space in literature and literature in space. In Perenič U (ed) Slavistična revija: Prostor v literaturi in literatura v prostoru/Space in Literature and Literature in Space, 60(3). pp 265–270

Schlosser HD (1987) dtv-Atlas zur deutschen Literatur. Tafeln und Texte. 3rd edn. Taschenbuch Verlag GmbH & Co. KG, München

Schmidt SJ (1980, 1991) Grundriss der Empirischen Literaturwissenschaft. Frankfurt/Main Suhrkamp

Perenič U (ed) (2012) Slavistična revija (Journal for Linguistics and Literary Studies), vol 60(3). Prostor v literaturi in literatura v prostoru/space in literature and literature in space (thematic issues). Ljubljana: Slavistično društvo Slovenije

Elements of Historical Knowledge About Urban Spaces: Reflections on the Requirements for a Dynamic Map

Ekkehard Schönherr

Abstract Various interpretations of Barcelona's urban space are presented in this paper in connection with the city's key structural changes during the early modern period. In line with certain utilitarian perspectives, the contemporary actors at various points in time came up with diverse models of spatial structurisation. In addition to the apparently natural division of Barcelona into two parts, an administrative division into four quarters had been established since the Middle Ages. However, this division was not directly linked to the pre-dominant, binary interpretation. Both in descriptions of the city by German speaking authors and in artistic depictions, binary structurisations of the city remained in the forefront. However, alongside the dominant perception of the city as being divided along the inner city wall and the Ramblas, alternative interpretations made their appearance, based on the old Roman city wall or the city's later extensions. Modern maps of historical space should not depict merely established and pre-dominant interpretations, but their task is also to present alternative and minority structurisations. In addition, there is an indication in this process of the danger which is inherent in the mapping of urban spaces: complete mapping of the historical city often gives the impression that the whole of the city space is covered by our knowledge. We are, however, confronted with huge informational gaps as far as the historical general image of the city is concerned. This is because generally only a select number of building types are depicted or described, or, alternatively, these depictions and descriptions refer only to a narrowly defined area. This information gap should also be recognized in maps portraying the historical city. Just as the representation of our knowledge is essential, it is also equally important to represent our information gaps.

E. Schönherr (✉)
Geschichte und Kulturen der Räume in der Frühen Neuzeit, University of Erfurt, Erfurt, Germany
e-mail: ekkeschoenherr@gmx.de

S. Rau and E. Schönherr (eds.), *Mapping Spatial Relations, Their Perceptions and Dynamics*, Lecture Notes in Geoinformation and Cartography,
DOI: 10.1007/978-3-319-00993-3_10,

1 Essential Features of the Urban Development of Barcelona, 1714–1854

During the 16th and 17th centuries there had been no radical changes in Barcelona's urban structure. Later historians had seen this as a symptom of the city's stagnation, or even of its decadence (see García Espuche 1998: 13–17). Only during the last two decades has this view of Barcelona undergone a change. Gabriel García i Espuche has shown how, in the decades around 1600, a new hierarchized urban structure was emerging in Catalonia, with Barcelona at the head. A key element among these changes was that a large proportion of the production was moved to the towns in the surrounding area, which then enjoyed a period of considerable growth with the consequence that Barcelona had an even more important role to play with regard to the marketing of various products (García Espuche 1998). Its economic boom in the 17th century was accompanied by numerous architectural changes, which at the micro-architectural level were manifested mainly in architectural fashions together with the adding of extensions to individual buildings; in addition, this process included not only the construction of numerous new commercial buildings but also the founding of a whole range of new religious houses (Perelló Ferrer 1996).[1] Except for this conventualisation, which mainly affected the Raval situated in the western part of the city, the main municipal structures were largely unaffected by these developments and remained in a basically unaltered condition from 1516 to 1716 (García Espuche 1998: 74–75).

Since the "Reapers' War" (cat.: Guerra dels Segadors 1640–1652) fighting for Catalonian independence, Barcelona had been frequently besieged (and sometimes for a very long time) up to 1714. The building expansion of the various fortifications during this period prefaced the dominant role with regard to municipal building which the military had played after the end of the last great Siege of Barcelona (September 1714). Catalonia's defeat in the Spanish Wars of Succession from 1701–1714 (cat.: Guerra de Successió Espanyola) signified the end to its special privileges and resulted in Catalonia's integration into the Spanish unitary state. A citadel was completed by 1719 in the north eastern area of the fortification sites in order to provide military security for Barcelona. This new fortress did not only cater for the external defence of the city but, together with the older fortress on the Montjuich and numerous barracks scattered over the city area, it was also responsible for controlling the city. A significant section of the northern part of Barcelona was demolished to make way for the construction of the Citadel, which had a deleterious effect, however, on the most economically active city centre (Busquets 2004: 85–88, see also the article by Guàrdia Bassols in this volume).

The construction of the Citadel marked the first great breach from Barcelona's medieval city structure. This change was initiated and maintained by the military as was the case with the two other major intrusions of the 18th century: firstly,

[1] In addition, the demographic trends analysed by Manel Guàrdia Bassols (see his contribution in the present volume), demonstrate the vitality of what had been assumed to be a stagnating city.

from 1753 the *extramuros* construction of the Barceloneta built on the alluvial sand next to the harbor and secondly, the demolition of the inner city wall from 1775 (Fig. 1).

In the second half of the 18th century Barcelona had to face the lasting problem of a rapidly growing population. It increased, between 1759 and 1787, from 53,000 to 95,000 inhabitants (Vilar 1973: 43–44). This growth obviously could not be accommodated by the new district of the Barceloneta. Because of the fortifications, the city could not expand outwards. So solutions had to be found inside the walls. The answer to the population explosion was the development of the Raval district, which up to that time was mainly used for agriculture. To promote this project, which began in 1775, the inner city wall was demolished and the Rambla was transformed into an avenue. This attempt was not very successful, but, ten years later, after the opening of the street Calle del Conde del Asalto, an intensive process of opening up new streets began to take place in the Raval. That was the moment at which the citizens of Barcelona representing their private interests once again began to play an active role in the city's development. By opening up new streets until 1808 (see Fig. 2), the character of the Raval changed completely and in the early 19th century, the quarter became the most important industrial and workers' district (Garcia i Espuche and Guàrdia i Bassols 1986: 72–74; Guàrdia et al. 1994: 70, 74).

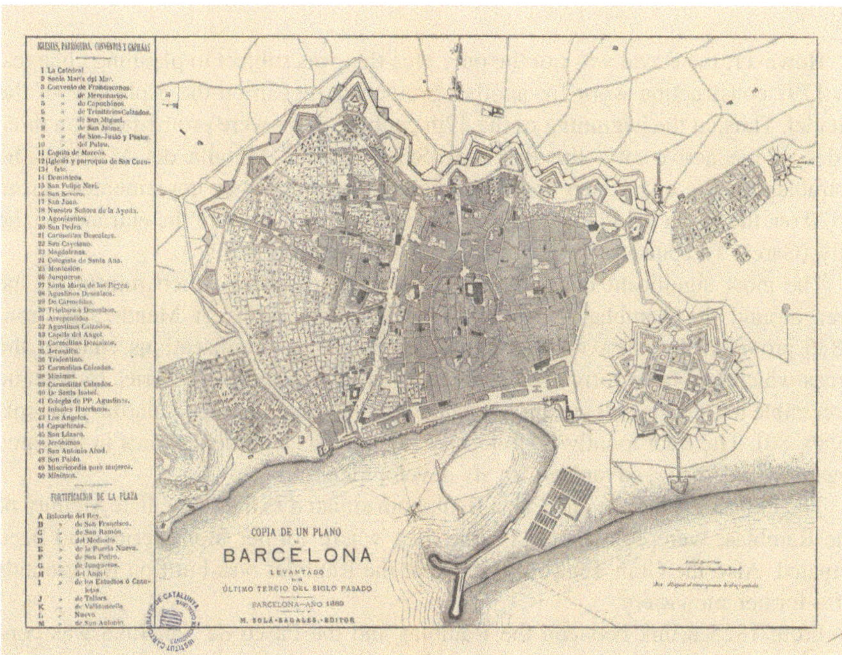

Fig. 1 M. Solá-Sagalés, Copia de un plano de Barcelona levantado en el último tercio del siglo pasado, 1889, (State approx. 1775) (© Institut Cartogràfic de Catalunya, RM.42027)

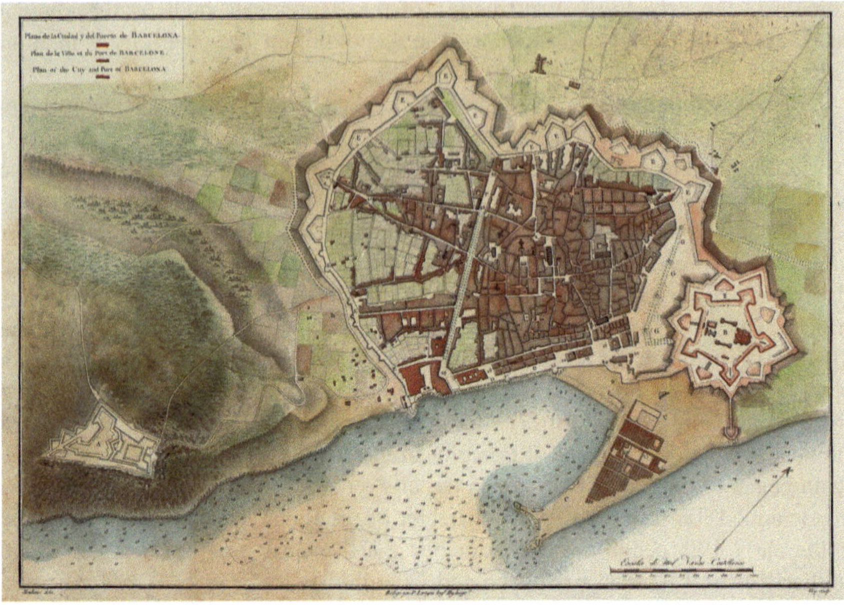

Fig. 2 J. Moulinier, Plano de la ciudad y del puerto de Barcelona, 1807 (© Institut Cartogràfic de Catalunya, RM.19425)

However, the Raval was not the only area that was subject to planning changes. In 1797 construction work for another promenade began on the esplanade of the citadel. Thus, at the beginning of the 19th century, there were promenades on three sides of the active city: the Paseo de San Juan, the Muralla del Mar and the Ramblas. In the city center a road-straightening policy was implemented. From 1820 on, the Plaza del Palacio was extended from the harbor to the entrance to the city (Garcia i Espuche and Guàrdia i Bassols 1986: Fig. 77).

The most significant changes, however, involved those which reinforced the area around the Ramblas. The ecclesiastical confiscations of Mendizabal from 1836 provided the basis for this reinforcement. These confiscations enabled the areas which had been formerly occupied by churches and monasteries to become accessible to the urban development again (Garcia i Espuche and Guàrdia i Bassols 1986: 80). The traditionally high density of churches and monasteries in this area was now a decisive advantage for its development.

The Plaza de San Jose and Plaza Real, both situated in the immediate vicinity of the Ramblas, were constructed on the area where former monasteries had been situated. Also the Gran Teatro del Liceo at the Ramblas was built on the grounds of a former monastery.

From 1825 a link between the Ramblas and the Paseo de San Juan was constructed in several stages. Also today's Plaza de San Jaime has been enlarged. In this connection, the new main axis formed by the Calle de Fernando VII and Calle

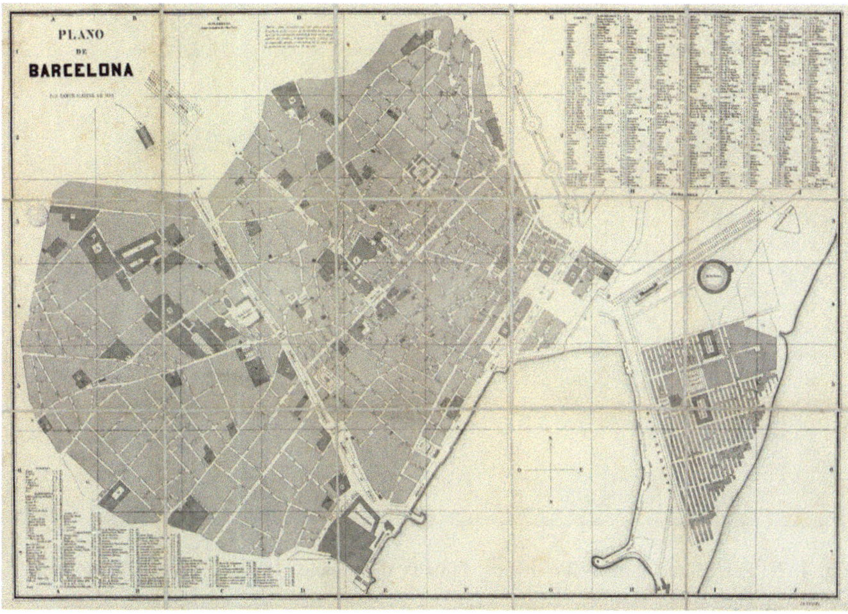

Fig. 3 R. Alabern, Plano de Barcelona, 1857 (© Institut Cartogràfic de Catalunya, RM.49231)

de la Princesa created a new urban cross which completed and, in parts, replaced the traditional inner city's structure, which up to then was still defined by the Roman *cardo* and *decumanus maximus* (Guàrdia et al. 1994: 74; see Fig. 3).

The most important buildings of political and economic administration remained. To maintain that the center was translocated would be an exaggeration as it was, in fact, only enlarged. However, visitors' perceptions of the city experienced fundamental changes. In this context, the Ramblas area became dominant.

2 Representations of Barcelona's Urban Space in Maps of the Early Modern Age

The first maps of Barcelona were made in the 17th century. Like the first city views, these maps are not indigenous productions, but commissioned works for outside entities such as monarchs or the military. Accordingly, the fortifications are particularly important in these maps, which, however, are very inaccurate. The fortifications are exaggerated and the roads bear little correspondence to reality (see Fig. 4, for the early cartography of Barcelona see the contributions in Montaner and Nadal 2010).

The relation between the Raval quarter and the genuine city is of particular interest. Although the Raval is mainly used for agriculture and is very sparsely

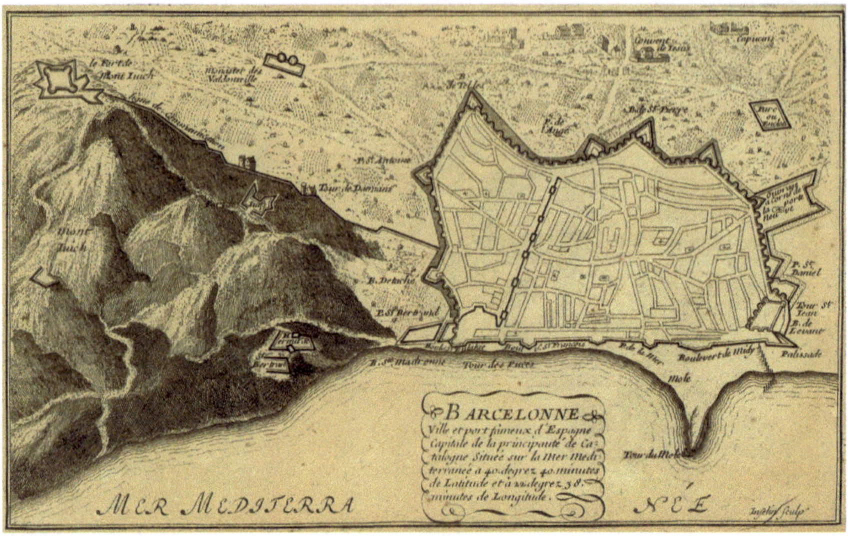

Fig. 4 N. Fer, Barcelonne. Ville et port fameux d'Espagne, 1690 (© Institut Cartogràfic de Catalunya, RM. 215332)

populated, both parts appear identical here. Little or nothing points to the fundamental differences between the two parts of the city and even the road network appears with a similar density.

In the context of the sieges between 1640 and 1714, a large number of maps were made of Barcelona in the decades around 1700. Here the technical and methodological improvements in the French cartography are visible. According to their military function, the main focus of these maps is again on the fortifications. Often, downtown areas are no longer shown.

Sometimes only the external system is represented. More often, however, the map shows the inner city wall, thus emphasizing the bi-partite division of Barcelona (see Fig. 5).

Frequently, the labels "old" and "new" were used for these two parts. Both are defining attributions that may be based on pragmatic decisions, but they are not very coherent. Given the completely different nature of the two parts, it is questionable whether the Raval is even part of the city although this question was apparently answered positively because of the existence of the closed city wall. However, the words "old" and "new" cease to apply as relevant terms if the wall is to be seen as being the defining element of the city.

This is because the construction of the wall around the Raval started after the construction of the wall around the densely populated city had begun, but this part was finished much sooner. The terms "old" and "new" are no longer a meaningful category for the entire system of fortifications. Moreover, the supposedly "new" town at the time of the making of these maps had been defined by the walls for over 200 years.

Fig. 5 Mortier, A True and Exact Plan of all the Old & New Fortifications of Barcelona, 1708 (© Institut Cartogràfic de Catalunya, RM.221454)

After the conquest by Castilian and French troops in 1714, military cartography experienced a boost of professionalization. Soon, a large number of standardized maps of the whole city were produced in the city itself showing its quarters and streets including items such as construction projects. In the context of military mapmaking during the second half of the 18th century, some maps appeared with representations of a new element, unknown until this moment in city maps of Barcelona. At a time of highly precise cartography, this new element, representing the Roman city walls, could be seen in maps, but not in reality because the Roman Wall had been integrated into downtown houses. Even this element was shown in these maps (Fig. 6).

Although the attributions "old" and "new" here are no longer explicitly present in the map, they are chronologically useful at this point. Since the Roman Wall did not exist as a continuity in the city's image, an interpretation is needed to explain why it was shown in the maps. It is not likely that this wall still had military importance at this time. Thus, its appearance may be understood as an expression of a different, more historically-determined interest in the city.

Fig. 6 Zermeño, Proyecto general para fortificar Barcelona, Ciudadela, Monjuyc y mejorar el Puerto, por D. Juan Zermeño en 31 Diciembre de 1751 (Copy from 1890, Detail) (© Arxiu Històric de la Ciutat de Barcelona, 02973)

As we shall see, the division of Barcelona with reference to the Roman city walls can also be found in some travelogues. It seems that at the beginning of the 19th century, this reference was no longer used in the maps of Barcelona, but, in the middle of the century, it appears again as a dominant element in Ildefonso Cerdà's and Miguel Garriga y Roca's plans for the inner city's reform. This time the function of this item is obvious. Beside the ideas of monument protection, it's easier not to come into contact with the city walls if you want to built a new broad road through the city center.

As these maps with the inscribed Roman walls were military maps, it is not very probable that they reached broader publication. However, in the 19th century, they could be copied and published, as we can see from Fig. 1, dated from 1889.

The perception of early modern Barcelona as a city comprising two parts which were divided by the inner city wall, seems to be immediately obvious. The maps focus on the walls which are crucial from a military perspective, but it can be assumed that these walls had an impact on the contemporaries' general perception of the city even in a non-military context. Nevertheless, the subdivision of the city on the lines of the medieval walls was neither natural nor necessary, but based on utilitarian principles representing particular interests. This can be seen both in the military maps which refer to the Roman Wall and in the division of the city's structure into its constituent administrative districts. Since the Middle Ages, the administration had divided Barcelona into four quarters by two main axes, which initially crossed at the Plaza del Angel and later at the Plaza de San Jaime (Brotons i Segarra 2008: 24). For this division, the city wall of the Ramblas was of little importance and, therefore, administratively more permeable than the main roads. It was not until 1770, thus shortly before the inner city wall had been pulled down, that the Raval was defined as an administratively independent quarter (Brotons i Segarra 2008: 69).

In the middle of the 19th century, it can also be seen that the improvement of cartographic techniques does not necessarily lead to objectification or depolitization of maps. In this example (Fig. 7), the placing of the map index just above the citadel is highly significant and can only be understood as a political statement. This symbolic extinction of the fortress anticipates its actual demolition by 11 years.

3 Barcelona's Urban Space in City Views and Travelogues

Although the early maps represented the city in a fairly unspecific way, they were at least an attempt to capture the city as a whole. This also applied to the first city views.

There are two well-known, very early examples. Both show the view of the city, taken from Montjuic, the hill beside Barcelona (Figs. 8 and 9).

The *veduta* of Wyngaerde is often singled out for praise for being extremely detailed and "of almost photographic fidelity" (Galera et al. 1982: 15). However,

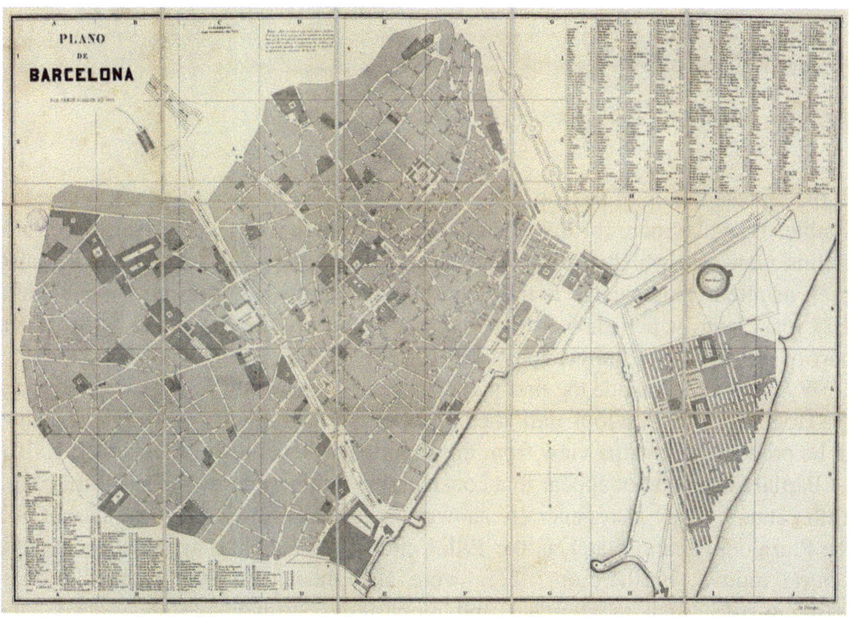

Fig. 7 R. Alabern, Plano de Barcelona, 1857 (© Institut Cartogràfic de Catalunya, RM.49231)

Fig. 8 J.C. Vermeyen/F. Hogenberg, Barcelona. Barcino, que vulgo Barcelona dicitur, 1572 (state of 1535) (© Institut Cartogràfic de Catalunya, RM.4165)

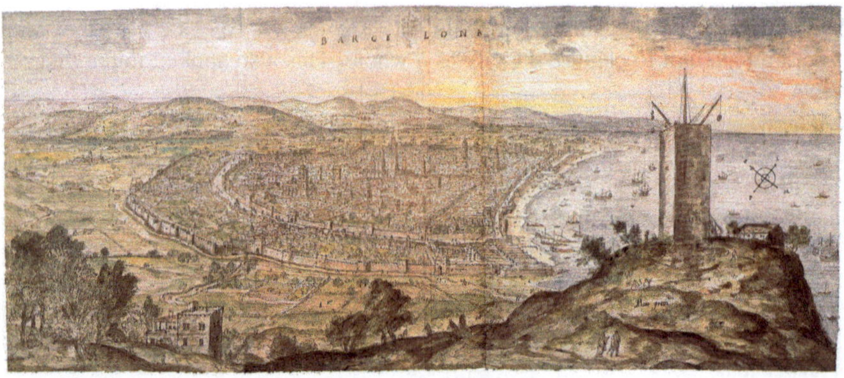

Fig. 9 A. van den Wyngaerde, Barcelona, 1563 (Wikimedia)

in the representation of the Raval the limits or the cost of such accuracy are clearly visible. To make room for the presentation of the actual city, the Raval has shrunk in this view, for which a significant deformation of the inner city wall has also been adopted. This change is the opposite case compared with the maps made a 100 years later, which showed the tendency to portray the Raval as a fully valued part of a relatively homogeneous city (Fig. 4).

Wyngaerde also made the first view from the seaside. After 80 years of copying the Hogenberg view, it took until the middle of the 17th century for the new *vedutas* to be produced. Now the view from the seaside has become the standard (Fig. 10).

Partial views do not appear in a considerable number before the last third of the 18th century. They also generally showed the sea side of the city, the harbor and the Plaza del Palacio. Later, in the 19th Century, the Ramblas appeared as another central motive. Furthermore, there were also thematically determined motifs: religious sites such as churches and monasteries from inside and outside, the administration buildings at the Plaza de San Jaime and the city palaces.

Fig. 10 Daumont, Barcelone, Ville capitale de la Principaute de Catalogne située sur la mer mediterranée, 1739–1748 (© Institut Cartogràfic de Catalunya, RM.215203)

Frequently, there are also depictions of rural scenes, with either a full or partial view of Barcelona in the background. But there are hardly any pictures of inner-city areas, which go beyond this framework—industrial areas for instance or poor quarters. Exceptions that do go beyond are mainly depictions of historical events. At a very early stage, there are representations of the storming of the city in 1714 and then, later depictions of urban unrest and riots. Finally, the demolition of the walls led to an increased production of pictures of the city wall as seen from outside (see Galera et al. 1982; CCCB/IMhB 1995).

The travel reports also provide only a very incomplete picture of the urban area. At least travelogues are the most complex type of source mentioned here, which also have a very long tradition. Since the Middle Ages there have been travelogues with references to Barcelona. The almost entirely positive description of the town is a striking feature of these reports, and Barcelona was often praised for its cleanliness and well-paved streets. These early reports offered only very brief accounts of the city and the most frequently discussed item was the port and its state. This is probably explained by the fact that travelers to Spain in the Middle and early modern Ages did not come across Barcelona, when they took the land route. When they took the sea route, they often went ashore here to continue to Toledo or Madrid. Due to the poor conditions of the harbor, this endeavour was for a long period of time both a complicated and newsworthy process.

Extensive descriptions of the city had been published in a greater density since the second half of the 18th century and had become numerous in the 19th century.

Now the ascent of Montjuic and a description of the entire city was a regular item on the traveler's program. Here the predominant representations were those which described the city as divided into two by the Ramblas. This was what the maps had also shown. However, over a long period of time, there are also perceptions where the Ramblas division does not play a role. These authors also referred to the Roman city walls, around which the medieval town had developed and to which the Barceloneta had been added.[2]

With the extension of the city, the downtown area is increasingly seen as a unit, which forms a stark contrast to the extension area. The Ramblas constitute the new center of this unit. Especially in the 19th century, they form the core of the reports on Barcelona. Beside the Ramblas, usually only the Muralla del Mar and the Calle del Fernando VII appear as a coherent space. Similar descriptions with a "space-like" reference refer perhaps to the Plaza Real, Plaza de la Constitución and Calle de Escudellers. Thus, the focus is on a continuous, newly-designed urban space. The descriptions of the port and the Barceloneta, which are often understood as a unit, are also quite large and numerous. There are also more localized descriptions, as we know them from the city views: churches (which generally also includes a description of the interior), monasteries, palaces and the seats of the municipal and provincial administration, the new cemetery, and also the Citadel. Even after the demolition of the Citadel, reports regularly referred to its being there on site. In addition, special interests are cultivated by the authors, who described, for example, hospitals, libraries or collections of paintings. It is noteworthy that the newly designed Paseo de San Juan is almost irrelevant—it is probably too remote. Also virtually no references are made to the newly developed area in the Raval, or, at most, in a summary manner, in the form of: "In recent decades, many new roads were opened here and there was a lot of construction." The old industrial area of San Pedro in the north of the city remained completely ignored.

4 Atypical Cartographic Representations

Up to this point I have presented relatively unambiguous spatial references from various sources, which refer to diverse functional interests and to individual features from a variety of text types. The situation becomes more complicated when

[2] Examples are: "It [Barcelona] is situated by the Mediterranean Sea and seems to be divided into two cities, one of which forms the inner city, which has high walls and four gates lead through these four walls corresponding to the four areas of the heavens and this part is called the Old Town. The other city is built around the Old Town and, likewise, has strong walls and solid towers and can be called the New Town." (M. 1704: 461); "Actually, it [Barcelona] consists of three cities: the Old Town and the New Town, and the Barceloneta with its harbour. The New Town almost comprises the whole city as it encompasses the Old Town, whose walls are still completely intact in many places. The building of these walls is often attributed to the Romans because the Spaniards, who know very little about architecture and the Ancient World, ascribe everything to the Romans" (Loning 1844: 21).

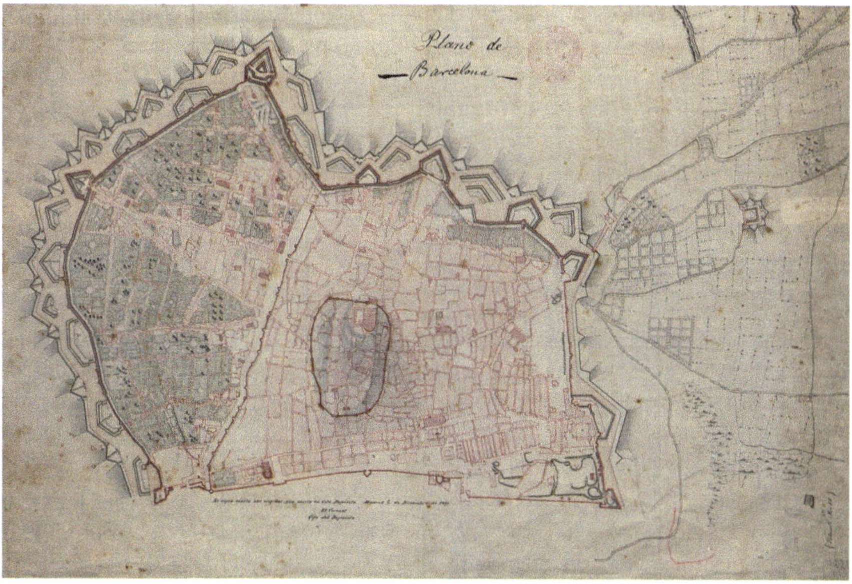

Fig. 11 Juan Zermeño (?), Plano de Barcelona—(Copy from 1891, Detail) (© Arxiu Històric de la Ciutat de Barcelona, 02551)

the concrete, historical map represents a complex network involving processes of map production and copying, anachronisms, outlines and even 'wishful thinking'.

Figure 11 displays an extremely interesting artifact as these parameters cross each other. The copy of the original map was produced in 1891 and was based on a military map of Barcelona drawn up around the middle of the 18th century.[3] Both the medieval inner city wall and the Roman Wall are clearly visible. The maker of this map can in all probability be ascribed to Juan Zermeño, who was both a military engineer and cartographer (see also Fig. 6). It can be seen from the divergent version of the Raval and the city centre that the city plan has still not been completed. However, the features which can be recognized from Barcelona's streets, are very exact (Galera et al. 1982: 205). The clearest anachronism is the representation of La Ribera as completely intact in the state it was before its partial demolition, and as it was before the construction of the Citadel, thus in its original state before the Siege of 1714. Alongside this, the extensions of the fortifications refer to a future version as this was after the time when the original map must have been produced. The southern link to the sea accessed by the *Puerta de la Madrona*

[3] The references to various 19th century copies of 18th century military maps in this article are based on my own work at the AHCB—*Arxiu Històric de la Ciutat de Barcelona*, (the city's administrative archives) to which the copies have been bequeathed. In addition, this also refers to Barcelona's interest in its own history in the context of the Catalonian *Renaixença,* which was crucial for the production of these copies.

town gate had never been implemented in this particular form. It is highly probable that this city plan was part of a draft drawing for an extension of the fortifications. In addition to this anomalous presentation of the past as manifested in the north and in the fortification plans referring to a future date in the south, an element of pure fiction had entered into the Raval area: the gardens laid out in a clearly baroque design are completely devoid of any historical foundation.

5 Conclusion

The maps of Barcelona presented here relate to the city in its totality. I have tried to show that they depicted clearly distinguishable images of the city, depending on the state of the art of cartography at the time and depending also on specific tasks and political interests. Overall, they tended to emphasize the urban dichotomy caused by the inner city wall and the Ramblas. In the 2nd half of the 18th century representations of the city appeared which showed the Roman city walls as another structuring element of the territory. We encounter this interpretation, over a longer period, even in isolated travelogues.

I referred to the administrative division into four as an alternative organization of space where the inner city wall was irrelevant. Pictures of the city provide more concrete images, which, however, only very late and then in a highly selective fashion also deal with inner-city areas.

The situation with travelogues is quite similar. It is noteworthy, but probably also not surprising, that similar spaces are described in travelogues and city views. In the field of pictorial representations, this is applicable especially to the harbor, the Ramblas and religious institutions, and thus to the churches and monasteries. Especially the travelogues of the mid-19th century preferred to describe areas that had been recently restructured such as the Ramblas together with the connecting streets, Fernando VII and Escudillers.

The question which arises, is: how can the differing spatial relationships, such as those we encounter in the different sources, be represented? Activities may include not only the actual urban development, but also the interpretations of urban space, which are manifested in its various representations. From the above, a few considerations can be derived:

1. Georeferencing is an important tool for a representation of items referring to space and for the comparisons of these items with each other. In addition, it is an essential criterion for assessing historical maps. However, in certain circumstances, it can become too constrained when creating new representations. Georeferenced maps of the whole city simulate a spatial objectivity, which fails to be true to most of the sources relevant for this mapping. This is particularly the case for travelogues, but it also applies even to maps and cityscapes, which (quite rightly) are highly rated on account of their exactness. The significant, most informative differences can be found just in these deviations and so they should be brought to light. There is no obligation to georeference.

2. Historical maps tend to allow for no ambiguities and thus often make urban data seem to be clearer than actually is the case. Generally, representations of the city as a whole do not correspond with individual experiences of the urban space. As can be seen from the travelogues, only a partial view of the city is represented. However, even here, there are some conventions, which can be represented. A representation can be tried by using cartograms, but then the blank areas would be lost. These unknown items also need to be representable. Strategies need to be developed to avoid a spurious clarity for those situations which are not clearly unambiguous. Alongside the "blank areas", there could be "dying data". This means data with a limited duration of validity. If there isn't new evidence, which either guarantees continuity or change, then this data could fade away with time before disappearing entirely.

References

All maps by The Cartographic Institute of Catalonia (ICC—Institut Cartogràfic de Catalunya), Historic Archive of the City of Barcelona (AHCB—Arxiu Històric de la Ciutat de Barcelona) except Fig. 9 (by Wikimedia)

Brotons i Segarra R (2008) La ciutat captiva. Barcelona 1714–1860. Albertí, Barcelona

Busquets J (2004) Barcelona. La construcción urbanística de una ciudad compacta. Ediciones del Serbal, Barcelona

Centre de Cultura Contemporània de Barcelona, Institut Municipal d'història de Barcelona (eds) (1995) Retrat de Barcelona. Barcelona, (cit.: CCCB/IMhB 1995)

Galera M, Roca F, Tarragó S (1982) Atlas de Barcelona. Segles XVI–XX. La Gaya Ciencia, Barcelona

Garcia i Espuche A, Guàrdia i Bassols M (1986) Espai i societat a la Barcelona pre-industrial. Edicions de la Magrana, Barcelona

García Espuche A (1998) Un siglo decisivo. Barcelona y Cataluña 1550–1640. Alianza, Barcelona

Guàrdia M, Monclús FJ, Oyón JL (1994) Atlas histórico de ciudades europeas. Salvat, Barcelona

Loning A (1844) Das spanische Volk in seinen Ständen, Sitten und Gebräuchen mit Episoden aus dem carlistischen Erbfolge Kriege. Hahn, Hannover

M. J (1704) Der Schau-Platz von Spanien und Portugall, auf welchem Zweyen Absätzen bey einem jeden vorläufig die Staats- und Kriegs-Geschichte und dann die Beschreibung fast aller darinnen befindlichen Festungen und Plätze aufgeführt, mit Contrefaiten und wahren Grund-Rissen annehmlicher gemacht werden: Dem curieusen Leser zur Beurtheilung der Affairen dieser Zeit sehr dienlich. Leuther, Amsterdam

Montaner C, Nadal F (eds) (2010) Aproximacions a la història de la cartografia de Barcelona. Barcelona

Perelló Ferrer AM (1996) L'arquitectura civil del segle XVII a Barcelona. Publicacions de l' Abadia de Montserrat, Barcelona

Vilar P (1973) Un moment critic en el creixement de Barcelona: 1774–1787. In: Vilar, Pierre. Assaigs sobre le Cataluña del segle XVIII, Curial, Barcelona, pp 43–55

Commentary

Wolfgang Spickermann

Abstract The fact that eHumanities together with digital Humanities can generally be put onto an equal footing with computational linguistics is reflected in the allocation pattern of institutional funding in Germany during the last few years. In line with this new direction, almost to the exclusion of all others, only those projects which are concerned with the assessment and analysis of texts in the very broadest context were being and still are being supported. Numerous representatives from the humanities such as the German Association of Historians (Verband der Historiker und Historikerinnen Deutschlands) have called for a radical re-think in their subjects and have highlighted the fact that eHumanities are by no means solely concerned with textual analysis.

The fact that eHumanities together with digital Humanities can generally be put onto an equal footing with computational linguistics is reflected in the allocation pattern of institutional funding in Germany during the last few years. In line with this new direction, almost to the exclusion of all others, only those projects which are concerned with the assessment and analysis of texts in the very broadest context were being and still are being supported. Numerous representatives from the humanities such as the German Association of Historians (Verband der Historiker und Historikerinnen Deutschlands) have called for a radical re-think in their subjects and have highlighted the fact that eHumanities are by no means solely concerned with textual analysis.

This is also the direction taken in the present volume. In this book, the enormous value (historical) maps have as a resource will be demonstrated by a mixture of theoretical argumentation and concrete case studies; it will also be shown how

W. Spickermann (✉)
Religionsgeschichte des Mittelmeerraumes in der römischen Antike,
Max-Weber-Center of the University of Erfurt, Erfurt, Germany
e-mail: wolfgang.spickermann@uni-erfurt.de

W. Spickermann
Institut für Alte Geschichte und Altertumskunde Karl-Franzens-Universität Graz,
Universitätsplatz 3, Graz 8010, Austria

S. Rau and E. Schönherr (eds.), *Mapping Spatial Relations, Their Perceptions and Dynamics*, Lecture Notes in Geoinformation and Cartography,
DOI: 10.1007/978-3-319-00993-3_11,

mapping can be linked to textual sources and finally, to what other uses intelligent GIS-systems can be applied outside the field of the historical sciences.

If the so-called "linguistic turn" and "cultural turn" have brought about an interdisciplinary awareness opening up humanities to linguistic and cultural studies, then this has manifested itself with the topics of time and **space**, which has further led to a very valuable new direction for the Geographical Sciences. In his contribution, Richard Rodger shows the enormous range of possibilities for historians which emerge from the "historical mapping" of Great Britain by using "open source mapping". Rodger defines this process as a "spatial turn". This approach opens up completely new perspectives for the use and evaluation of well-known source material by setting this data into a direct, spatial context. Amongst other things, his project "Visualising Urban Geographies (VUG)" has presented a GIS-based social history of the Edinburgh of the 19th and 20th centuries entitled "Mapping Edinburgh's Social History (MESH)" which impressively demonstrates the advantages to be gained (not only for the humanities) from combining digitalized source materials with GIS-based map layers.

During the last few years, the introduction of WEB 2.0 together with the possibility of working alongside voluntary data suppliers has led to a rapid increase of geo-data in the Internet. Projects such as Open Street Map or Wikipedia are impressive results of this "crowd sourcing". However, for the academics, the use of this kind of data constantly gives rise to questions of quality and reliability and, on the other hand, there is a need to be able to bring together data gathered from a variety of heterogeneous web sources onto one platform. In this context, Leif Scheuermann's contribution shows how the Webbletechnology developed in Japan is extremely useful as a platform for the combination of the most diverse data systems and as a technology for the cross-referencing of digital analytical methods. For this process, the so-called webbles (web-based lifelike entities) can be introduced and then developed. Webble technology is based on the idea that digitalized visual objects can be "crossed" with one another and thus produce functionally viable new objects so that this creates the ideal platform for the combination of diverse analytical procedures. Webbles enable users to process and distribute encapsulated or—"wrapped"—existing data resources and to combine individual media objects by the direct manipulation of processes such as "drag", "drop", "copy", and "paste", so that new objects can be combined and created without having to possess any special programming skills. Scheuermann also describes meta-data models and ontologies which allow for a more or less automated quality management. As a technological platform, Webble has opened up a wide range of possibilities to the scientific community, which is why this technology should be developed further on account of its applicability to eHumanities. For this wide-ranging enterprise, the "Interdisciplinary Center of eHumanities in History and Social Sciences (ICE)" was founded at the Max-Weber-Center of the University of Erfurt in order to become a research association involving academics from various universities and research centers.

In the first instance, maps and atlases provide the visual means for presenting geographical information. On the one hand, they serve to store, organize and construct geographical data, but this medium can also give visual form to more geographically locatable information taken from various fields such as history, politics and economics. Traditionally, maps have always been purely static objects; however, a dynamic visualization of processes depending on periods in time is an absolute desideratum not only for historical research but also for research purposes in both the social and poli-tical sciences. A variety of serially-based data covering longer periods of time obtained from a wide range of similar spatial sources is needed for a better understanding of the processes of long duration in historical societies, which cannot be adequately portrayed merely in a linear-based statistical evaluation (as in time-lines). By allocating a specific period of time to each object, a great variety of processes can be visualized in such a way that the numerous connections stretching over a long period of time can become both visible and tangible. The more recent GIS-based technologies are attempting to follow these paths even though these routes cannot always be pursued systematically as is, however, the goal of the "Adaptive Interactive Dynamic Atlas (AIDA)", (which is also being developed at the ICE in Erfurt). These technologies are also being used to be able to developed pedestrian movement in the ceremonial centers of the Maya culture in Honduras at various periods in time (Shawn G. Morton, Meaghan M. Peuramaki-Brown, Peter C. Dawson, and Jeffrey D. Seibert); and working within a space–time relationship, they have been impressively applied to the urban development project carried out by Benjamin N. Vis of the University of Leeds.

The case study dealing with Barcelona's urban development over a long period of time demonstrates in a most impressive way the advantages of a presentation of geographical sources covering the demographic, economic and political development of the city from the 15th to the 19th century as in the analysis carried out by Manel Guardia Bassols and Sergi Garriga. On the one hand, only in this way the dynamic processes of urban development can be given a visual form and thus allow for 'presentability', but, on the other hand, it can also be shown how people (the actors) perceived their surroundings at various points in time. The contribution on the urban development of Lyons (Bernard Gauthiez, Olivier Zeller) in the 18th and 19th centuries is on similar lines, but covers a shorter 'distance' in time. Here, with the aid of a database of dated archive materials such as tax registers, census materials, planning database applications, ownership changes etc., GIS-based mapping layers can be generated for any chosen point in time within the whole area of the city so that the social, architectural and political changes in the city can be rendered visually. These case studies are able to demonstrate the enormous advantages dynamic maps produced from databases have over what up till now have been the usual static theme-based maps.

Ekkehard Schönherr's contribution shows that it is absolutely essential for this enterprise to work out generally accepted standards for dynamic maps. Maps are basically dependent on their respective databases. The source material needs to be sufficiently informative and to be suitably adaptable to a proper database so that it can be visualized. It goes without saying that there should be a spatial component (or in other words, be locatable) and finally, there must be enough material to provide sufficient relevantly informative material. The two case studies referred to document what a daunting task it is to work through huge quantities of materials, but, in that case, the long-term gain is much greater. Similarly, the example of Barcelona (Susanne Rau) shows that, whereas the two models of urban history just referred to result in clearly definable space concepts and structures, this clarity no longer applies when movement in space together with the perception of space and its concomitant changes are taken into consideration. All this implies that particular demands are now being made on processing data for dynamic GIS-based maps, all of which still needs to be validated in concrete projects.

Maps are always interpretations of a space based on subjective experience. The GIS technology leads to the immediate supposition that an apparently unquestionable geometrically and mathematically objective analysis needs to take place. However, Stephen Read's critical essay shows that the boundaries between "subjective" perceivable space and quantifiable space can be absolutely fluid and that our perceptions of our surroundings can, in turn, be influenced by the application of the new technologies.

To return to the remarks made at the beginning of this commentary: even the literary analyses undertaken by the classical DigitalHumanities can, in certain cases, have a spatial dimension and be given visual form in a time–space relationship. A combination of data mining with geographical or cartographical information systems and network analysis system is still, however, a desideratum for future research. The first beginnings in this direction can be seen Stanford University's project "Republic of Letters" (http://republicofletters.stanford.edu/), which, together with other institutions, is working in co-operation with the University of Erfurt's Research Centre based in Gotha. Urška Perenič's contribution offers another concrete example of using cartographical representation and analysis to present the biographies of Slovenian writers.

All the contributions in this volume clearly demonstrate the enormous scope of database-supported dynamic maps using GIS-technologies. If, in the future, a great number of users who may lack background knowledge in this direction should succeed in being able to manipulate this technology, perhaps via the Webble platform, so that they could adapt the information made available in this way to their specific research interests, then this would really be a tremendous leap into the future. As long as information can be mappable, the encyclopedia of the future lies with maps that can be generated dynamically and on which the whole the relevant background material is available for both free linkage and visualizations.

MIX
Papier aus verantwortungsvollen Quellen
Paper from responsible sources
FSC® C105338

If you have any concerns about our products,
you can contact us on
ProductSafety@springernature.com

In case Publisher is established outside the EU,
the EU authorized representative is:
Springer Nature Customer Service Center GmbH
Europaplatz 3, 69115 Heidelberg, Germany

Printed by Libri Plureos GmbH
in Hamburg, Germany